CONCEPTS IN NUMBER THEORY

CONCEPTS IN NUMBER THEORY

Amit Kumar

Oxford Book Company
Jaipur ▪ New Delhi

Published By

Oxford Book Company

Regd. Off. "Bony Residency", Gate No. 2
Opp. Tilak Public School, Vishweswariya Nagar,
Gopalpura Road, Jaipur - 302018, India
Ph: 0141 - 2761280, Telefax: 2761381
email - oxfordbookjpr@gmail.com
website - oxfordbookcompany.com

Delhi Office: 102, 1st Floor, Satyam House,
4327/3, Ansari Road, Darya Ganj,
New Delhi - 110002, India
Ph: 011-45652440

Distributed By
KEYA INTERNATIONAL PRESS LLP
E-mail: keyainternationalpressllp@gmail.com

Published in India, 2025

ISBN : 9789391096472

Typeset by Shivangi Computers, Jaipur

Printed and bound in India by Thomson Press Ltd.,
New Delhi

Contents

Preface

Number theory is a branch of pure mathematics devoted primarily to the study of the integers. Number theorists study prime numbers as well as the properties of objects made out of integers (such as rational numbers) or defined as generalizations of the integers. Integers can be considered either in themselves or as solutions to equations (diophantine geometry). Questions in number theory are often best understood through the study of analytical objects (e.g., the Riemann zeta function) that encode properties of the integers, primes or other number-theoretic objects in some fashion (analytic number theory). One may also study real numbers in relation to rational numbers, e.g., as approximated by the latter (diophantine approximation). The older term for number theory is *arithmetic*. By the early twentieth century, it had been superseded by "number theory". The use of the term *arithmetic* for *number theory* regained some ground in the second half of the 20th century, arguably in part due to French influence. In particular, *arithmetical* is preferred as an adjective to *number-theoretic. While* Babylonian number theory—or what survives of Babylonian mathematics that can be called thus—consists of this single, striking fragment, Babylonian algebra (in the secondary-school sense of "algebra") was exceptionally well developed. Late Neoplatonic sources state that Pythagoraslearned mathematics from the Babylonians. Euclid IX 21—34 is very probably Pythagorean; it is very simple material ("odd times even is even", "if an odd number measures [= divides] an even number, then it also measures [= divides] half of it"), but it is all that is needed to prove that $\sqrt{2}$ is irrational. Pythagoraean mystics gave great importance to the odd and the even.

The discovery that is irrational is credited to the early Pythagoreans (pre-Theodorus). By revealing (in modern terms) that numbers could be irrational, this discovery seems to have provoked the first foundational crisis in mathematical history; its proof or its divulgation are sometimes credited to Hippasus, who was expelled or split from the Pythagorean sect. It is only here that we can start to speak of a clear, conscious division between *numbers* (integers and the rationals—the subjects of arithmetic) and *lengths* (real numbers, whether rational or not). The Pythagorean tradition spoke also of so-called polygonal or figured numbers. While square numbers, cubic numbers, etc., are seen now as more natural than triangular numbers, square numbers, pentagonal numbers, etc., the study of the sums of triangular and pentagonal numbers would prove fruitful in the early modern period (17th to early 19th century). We know of no clearly arithmetical material in ancient Egyptian or Vedic sources, though there is some algebra in both. The Chinese remainder theorem appears as an exercise in Sun Zi's*Suan Ching* (also known as *The Mathematical Classic of Sun Zi*. There is also some numerical mysticism in Chinese mathematics, but, unlike that of the Pythagoreans, it seems to have led nowhere. Like the Pythagoreans' perfect numbers, magic squares have passed from superstition into recreation. Aside from a few fragments, the mathematics of Classical Greece is known to us either through the reports of contemporary non-mathematicians or through mathematical works from the early Hellenistic period. In the case of number theory, this means, by and large, *Plato* and *Euclid,* respectively. Plato had a keen interest in mathematics, and distinguished clearly between arithmetic and calculation. (By *arithmetic* he meant, in part, theorising on number, rather than what *arithmetic* or*number theory* have come to mean.) It is through one of Plato's dialogues—namely, *Theaetetus* – that we know that Theodorus had proven that are irrational. Theaetetus was, like Plato, a disciple of Theodorus's; he worked on distinguishing different kinds of incommensurables, and was thus arguably a pioneer in the study of number systems.

The book is replete with features which enable the building of a firm foundation of the underlying principles of the subject and also provide adequate scope for testing the comprehension acquired by the students.

—Amit Kumar

1

Introduction

Number Theory

Number theory (or arithmetic) is a branch of pure mathematics devoted primarily to the study of the integers. Number theorists study prime numbers as well as the properties of objects made out of integers (such as rational numbers) or defined as generalizations of the integers (such as, for example, algebraic integers).

Figure: *A Lehmer sieve - a primitive digital computer once used for finding primes and solving simple diophantine equations.*

Integers can be considered either in themselves or as solutions to equations (diophantine geometry). Questions in number theory are often best understood through the study of analytical objects (e.g., the Riemann zeta function) that encode properties of the integers, primes or other number-theoretic objects in some fashion (analytic number theory). One may also study real numbers in relation to rational numbers, e.g., as approximated by the latter (diophantine approximation).

The older term for number theory is *arithmetic*. By the early twentieth century, it had been superseded by "number theory". (The word "arithmetic" is used by the general public to mean "elementary calculations"; it has also acquired other meanings in mathematical logic, as in *Peano arithmetic*, and computer science, as in *floating point arithmetic*.) The use of the term *arithmetic* for *number theory* regained some ground in the second half of the 20th century, arguably in part due to French influence. In particular, *arithmetical* is preferred as an adjective to *number-theoretic*.

History

Origins

Dawn of arithmetic: The first historical find of an arithmetical nature is a fragment of a table: the broken clay tablet Plimpton 322 (Larsa, Mesopotamia, ca. 1800 BCE) contains a list of "Pythagorean triples", i.e., integers (a,b,c) such that $a^2+b^2=c^2$. The triples are too many and too large to have been obtained by brute force. The heading over the first column reads: "The *takiltum* of the diagonal which has been substracted such that the width..."

Figure: The Plimpton 322 tablet

The table's layout suggests that it was constructed by means of what amounts, in modern language, to the identity

$$\left(\frac{1}{2}\left(x-\frac{1}{x}\right)\right)^2+1=\left(\frac{1}{2}\left(x+\frac{1}{x}\right)\right)^2,$$

which is implicit in routine Old Babylonian exercises. If some other method was used, the triples were first constructed and then reordered by c/a, presumably for actual use as a "table", i.e., with a view to applications.

We do not know what these applications may have been, or whether there could have been any; Babylonian astronomy, for example, truly flowered only later. It has been suggested instead that the table was a source of numerical examples for school problems.

While Babylonian number theory—or what survives of Babylonian mathematics that can be called thus—consists of this single, striking fragment, Babylonian algebra (in the secondary-school sense of "algebra") was exceptionally well developed. Late Neoplatonic sources state that Pythagoras learned mathematics from the Babylonians. (Much earlier sources state that Thales and Pythagoras travelled and studied in Egypt.)

Euclid IX 21—34 is very probably Pythagorean; it is very simple material ("odd times even is even", "if an odd number measures [= divides] an even number, then it also measures [= divides] half of it"), but it is all that is needed to prove that $\sqrt{2}$ is irrational. Pythagoraean mystics gave great importance to the odd and the even.

The discovery that $\sqrt{2}$ is irrational is credited to the early Pythagoreans (pre-Theodorus). By revealing (in modern terms) that numbers could be irrational, this discovery seems to have provoked the first foundational crisis in mathematical history; its proof or its divulgation are sometimes credited to Hippasus, who was expelled or split from the Pythagorean sect. It is only here that we can start to speak of a clear, conscious division between *numbers* (integers and the rationals—the subjects of arithmetic) and *lengths* (real numbers, whether rational or not).

The Pythagorean tradition spoke also of so-called polygonal or figured numbers. While square numbers, cubic numbers, etc., are seen now as more natural than triangular numbers, square numbers, pentagonal numbers, etc., the study of the sums of triangular and pentagonal numbers would prove fruitful in the early modern period (17th to early 19th century).

We know of no clearly arithmetical material in ancient Egyptian or Vedic sources, though there is some algebra in both. The Chinese remainder theorem appears as an exercise in Sun Zi's *Suan Ching* (also known as *The Mathematical Classic of Sun Zi* (3rd, 4th or 5th century CE.) (There is one important step glossed over in Sun Zi's solution: it is the problem that was later solved by Âryabhama's kummaka.)

There is also some numerical mysticism in Chinese mathematics, but, unlike that of the Pythagoreans, it seems to have led nowhere. Like the Pythagoreans' perfect numbers, magic squares have passed from superstition into recreation.

Classical Greece and the Early Hellenistic Period

Aside from a few fragments, the mathematics of Classical Greece is known to us either through the reports of contemporary non-mathematicians or through mathematical works from the early Hellenistic period. In the case of number theory, this means, by and large, *Plato* and *Euclid*, respectively.

Plato had a keen interest in mathematics, and distinguished clearly between arithmetic and calculation. (By *arithmetic* he meant, in part, theorising on number, rather than what *arithmetic* or *number theory* have come to mean.) It is through one of Plato's dialogues—namely, *Theaetetus* – that we know that Theodorus had proven that $\sqrt{3}, \sqrt{5}, \ldots, \sqrt{17}$ are irrational. Theaetetus was, like Plato, a disciple of Theodorus's; he worked on distinguishing different kinds of incommensurables, and was thus arguably a pioneer in the study of number systems.

Euclid devoted part of his *Elements* to prime numbers and divisibility, topics that belong unambiguously to number theory and are basic thereto. In particular, he gave an algorithm for computing the greatest common divisor of two numbers (the

Euclidean algorithm; *Elements*, Prop. VII.2) and the first known proof of the infinitude of primes (*Elements*, Prop. IX.20).

In 1773, Lessing published an epigram he had found in a manuscript during his work as a librarian; it claimed to be a letter sent by Archimedes to Eratosthenes. The epigram proposed what has become known as Archimedes' cattle problem; its solution (absent from the manuscript) requires solving an indeterminate quadratic equation (which reduces to what would later be misnamed Pell's equation). As far as we know, such equations were first successfully treated by the Indian school. It is not known whether Archimedes himself had a method of solution.

Diophantus

Very little is known about Diophantus of Alexandria; he probably lived in the third century CE, that is, about five hundred years after Euclid. Six out of the thirteen books of Diophantus's *Arithmetica* survive in the original Greek; four more books survive in an Arabic translation. The *Arithmetica* is a collection of worked-out problems where the task is invariably to find rational solutions to a system of polynomial equations, usually of the form $f(x,y)=z^2$ or $f(x,y,z)=w^2$. Thus, nowadays, we speak of *Diophantine equations* when we speak of polynomial equations to which rational or integer solutions must be found.

One may say that Diophantus was studying rational points — i.e., points whose coordinates are rational — on curves and algebraic varieties; however, unlike the Greeks of the Classical period, who did what we would now call basic algebra in geometrical terms, Diophantus did what we would now call basic algebraic geometry in purely algebraic terms. In modern language, what Diophantus does is to find rational parametrisations of many varieties; in other words, he shows how to obtain infinitely many rational numbers satisfying a system of equations by giving a procedure that can be made into an algebraic expression (say, $x=g_1(r,s)$, $y=g_2(r,s)$, $z=g_3(r,s)$, where g_1, g_2 and g_3 are polynomials or quotients of polynomials; this would be what is sought for if such x,y,z satisfied a given equation $f(x,y)=z^2$ (say) for all values of r and s).

Diophantus also studies the equations of some non-rational curves, for which no rational parametrisation is possible. He manages to find some rational points on these curves – elliptic curves, as it happens, in what seems to be their first known occurrence—by means of what amounts to a tangent construction: translated into coordinate geometry (which did not exist in Diophantus's time), his method would be visualised as drawing a tangent to a curve at a known rational point, and then finding the other point of intersection of the tangent with the curve; that other point is a new rational point. (Diophantus also resorts to what could be called a special case of a secant construction.)

While Diophantus is concerned largely with rational solutions, he assumes some results on integer numbers; in particular, he seems to assume that every integer is the sum of four squares, though he never states as much explicitly.

Indian School: Âryabhama, Brahmagupta, Bhâskara

While Greek astronomy—thanks to Alexander's conquests—probably influenced Indian learning, to the point of introducing trigonometry, it seems to be the case that Indian mathematics is otherwise an indigenous tradition; in particular, there is no evidence that Euclid's Elements reached India before the 18th century.

Âryabhama (476–550 CE) showed that pairs of simultaneous congruences $n \equiv a_1 \pmod{m}_1$, $n \equiv a_2 \pmod{m}_2$ could be solved by a method he called *kummaka,* or *pulveriser;* this is a procedure close to (a generalisation of) the Euclidean algorithm, which was probably discovered independently in India. Âryabhama seems to have had in mind applications to astronomical calculations.

Brahmagupta (628 CE) started the systematic study of indefinite quadratic equations—in particular, the misnamed Pell equation, in which Archimedes may have first been interested, and which did not start to be solved in the West until the time of Fermat and Euler. Later Sanskrit authors would follow, using Brahmagupta's technical terminology. A general procedure (the chakravala, or "cyclic method") for solving Pell's equation was finally found by Jayadeva (cited in the eleventh century; his work is otherwise lost); the earliest surviving exposition appears in Bhâskara II's BîjagaGita (twelfth century). Unfortunately, Indian mathematics remained largely unknown in the West until the late eighteenth century; Brahmagupta and Bhâskara's work was translated into English in 1817 by Henry Colebrooke.

Arithmetic in the Islamic Golden Age

Figure: *Al-Haytham seen by the West: frontispice of* Selenographia, *showing Alhasen [sic] representing knowledge through reason, and Galileo representing knowledge through the senses.*

In the early ninth century, the caliph Al-Ma'mun ordered translations of many Greek mathematical works and at least one Sanskrit work (the *Sindhind*, which may or may not be Brahmagupta's Brâhmasphuþasiddhânta), thus giving rise to the rich tradition of Islamic mathematics.

Diophantus's main work, the *Arithmetica*, was translated into Arabic by Qusta ibn Luqa (820–912). Part of the treatise *al-Fakhri* (by al-Karajî, 953 – ca. 1029) builds on it to some extent. According to Rashed Roshdi, Al-Karajî's contemporary Ibn al-Haytham knew what would later be called Wilson's theorem.

Other than a treatise on squares in arithmetic progression by Fibonacci – who lived and studied in north Africa and Constantinople during his formative years, ca. 1175–1200 – no number theory to speak of was done in western Europe during the Middle Ages.

Matters started to change in Europe in the late Renaissance, thanks to a renewed study of the works of Greek antiquity. A key catalyst was the textual emendation and translation into Latin of Diophantus's *Arithmetica* (Bachet, 1621, following a first attempt by Xylander, 1575).

Early Modern Number Theory

Fermat: Pierre de Fermat (1601–1665) never published his writings; in particular, his work on number theory is contained almost entirely in letters to mathematicians and in private marginal notes. He wrote down nearly no proofs in number theory; he had no models in the area. He did make repeated use of mathematical induction, introducing the method of infinite descent.

One of Fermat's first interests was perfect numbers (which appear in Euclid, *Elements* IX) and amicable numbers; this led him to work on integer divisors, which were from the beginning among the subjects of the correspondence (1636 onwards) that put him in touch with the mathematical community of the day. He had already studied Bachet's edition of Diophantus carefully; by 1643, his interests had shifted largely to diophantine problems and sums of squares (also treated by Diophantus).

Fermat's achievements in arithmetic include:

- Fermat's little theorem (1640), stating that, if a is not divisible by a prime p, then $a^{p-1} \equiv 1 \pmod{p}$.
- If a and b are coprime, then $a^2 + b^2$ is not divisible by any prime congruent to -1 modulo 4; *and* Every prime congruent to 1 modulo 4 can be written in the form $a^2 + b^2$. These two statements also date from 1640; in 1659, Fermat stated to Huygens that he had proven the latter statement by the method of descent. Fermat and Frenicle also did some work (some of it erroneous or non-rigorous) on other quadratic forms.

- Fermat posed the problem of solving $x^2 - Ny^2 = 1$ as a challenge to English mathematicians (1657). The problem was solved in a few months by Wallis and Brouncker. Fermat considered their solution valid, but pointed out they had provided an algorithm without a proof (as had Jayadeva and Bhaskara, though Fermat would never know this.) He states that a proof can be found by descent.
- Fermat developed methods for (doing what in our terms amounts to) finding points on curves of genus 0 and 1. As in Diophantus, there are many special procedures and what amounts to a tangent construction, but no use of a secant construction.
- Fermat states and proves (by descent) to *Observations on Diophantus* (Obs. XLV) that $x^4 + y^4 = z^4$ has no non-trivial solutions in the integers. Fermat also mentioned to his correspondents that $x^3 + y^3 = z^3$ has no non-trivial solutions, and that this could be proven by descent. The first known proof is due to Euler (1753; indeed by descent).

Fermat's claim ("Fermat's last theorem") to have shown there are no solutions to $x^n + y^n = z^n$ for all $n \geq 3$ (a fact completely beyond his methods) appears only on his annotations on the margin of his copy of Diophantus; he never claimed this to others and thus had no need to retract it if he found a mistake in his alleged proof.

Euler

The interest of Leonhard Euler (1707–1783) in number theory was first spurred in 1729, when a friend of his, the amateur Goldbach pointed him towards some of Fermat's work on the subject. This has been called the "rebirth" of modern number theory, after Fermat's relative lack of success in getting his contemporaries' attention for the subject. Euler's work on number theory includes the following:

- *Proofs for Fermat's statements.* This includes Fermat's little theorem (generalised by Euler to non-prime moduli); the fact that $p = x^2 + y^2$ if and only if $p \equiv 1 \; mod \; 4$; initial work towards a proof that every integer is the sum of four squares (the first complete proof is by Lagrange (1770), soon improved by Euler himself; the lack of non-zero integer solutions to $x^4 + y^4 = z^2$ (implying the case *n*=4 of Fermat's last theorem, the case *n*=3 of which Euler also proved by a related method).

- *Pell's equation,* first misnamed by Euler. He wrote on the link between continued fractions and Pell's equation.
- *First steps towards analytic number theory.* In his work of sums of four squares, partitions, pentagonal numbers, and the distribution of prime numbers, Euler pioneered the use of what can be seen as analysis (in particular, infinite series) in number theory. Since he lived before the development of complex analysis, most of his work is restricted to the formal manipulation of power series. He did, however, do some very notable (though not fully rigorous) early work on what would later be called the Riemann zeta function.
- *Quadratic forms.* Following Fermat's lead, Euler did further research on the question of which primes are can be expressed in the form $x^2 + Ny^2$, some of it prefiguring quadratic reciprocity.
- *Diophantine equations.* Euler worked on some diophantine equations of genus 0 and 1. In particular, he studied Diophantus's work; he tried to systematise it, but the time was not yet ripe for such an endeavour – algebraic geometry was still in its infancy. He did notice there was a connection between diophantine problems and elliptic integrals, whose study he had himself initiated.

Lagrange, Legendre and Gauss

Lagrange (1736–1813) was the first to give full proofs of some of Fermat's and Euler's work and observations - for instance, the four-square theorem and the basic theory of the misnamed "Pell's equation" (for which an algorithmic solution was found by Fermat and his contemporaries, and also by Jayadeva and Bhaskara II before them.) He also studied quadratic forms in full generality (as opposed to $mX^2 + nY^2$) — defining their equivalence relation, showing how to put them in reduced form, etc. Legendre (1752–1833) was the first to state the law of quadratic reciprocity. He also conjectured what amounts to the prime number theorem and Dirichlet's theorem on arithmetic progressions. He gave a full treatment of the equation $ax^2 + by^2 + cz^2 = 0$ and worked on quadratic forms along the lines later developed fully by Gauss. In his old age, he was the first to prove "Fermat's last theorem" for $n = 5$ (completing work by Dirichlet, and crediting both him and Sophie Germain).

In his *Disquisitiones Arithmeticae* (1798), Gauss (1777–1855) proved the law of quadratic reciprocity and developed the theory of quadratic forms (in particular, defining their composition). He also introduced some basic notation (congruences) and devoted a section to computational matters, including primality tests. The last section of the *Disquisitiones* established a link between roots of unity and number theory:

> The theory of the division of the circle...which is treated in sec. 7 does not belong by itself to arithmetic, but its principles can only be drawn from higher arithmetic.

In this way, Gauss arguably made a first foray towards both Galois's work and algebraic number theory.

Maturity and Division into Subfields

Starting early in the nineteenth century, the following developments gradually took place:

- The rise to self-consciousness of number theory (or *higher arithmetic*) as a field of study.
- The development of much of modern mathematics necessary for basic modern number theory: complex analysis, group theory, Galois theory—accompanied by greater rigor in analysis and abstraction in algebra.
- The rough subdivision of number theory into its modern subfields—in particular, analytic and algebraic number theory.

Algebraic number theory may be said to start with the study of reciprocity and cyclotomy, but truly came into its own with the development of abstract algebra and early ideal theory and valuation theory. A conventional starting point for analytic number theory is Dirichlet's theorem on arithmetic progressions (1837), whose proof introduced L-functions and involved some asymptotic analysis and a limiting process on a real variable. The first use of analytic ideas in number theory actually goes back to Euler (1730s), who used formal power series and non-rigorous (or implicit) limiting arguments.

The use of *complex* analysis in number theory comes later: the work of Riemann (1859) on the zeta function is the canonical starting point; Jacobi's four-square theorem (1839), which predates it, belongs to an initially different strand that has by now taken a leading role in analytic number theory (modular forms).

The history of each subfield is briefly addressed in its own section below; see the main article of each subfield for fuller treatments. Many of the most interesting questions in each area remain open and are being actively worked on.

Main Subdivisions

Elementary tools: The term *elementary* generally denotes a method that does not use complex analysis. For example, the prime number theorem was first proven in 1896, but an elementary proof was found only in 1949 by Erdõs and Selberg. The term is somewhat ambiguous: for example, proofs based on complex Tauberian theorems (e.g. Wiener–Ikehara) are often seen as quite enlightening but not elementary, in spite of using Fourier analysis, rather than complex analysis as such. Here as elsewhere, an *elementary* proof may be longer and more difficult for most readers than a non-elementary one.

Number theory has the reputation of being a field many of whose results can be stated to the layperson. At the same time, the proofs of these results are not particularly accessible, in part because the range of tools they use is, if anything, unusually broad within mathematics.

Analytic Number Theory

In mathematics, analytic number theory is a branch of number theory that uses methods from mathematical analysis to solve problems about the integers. It is often said to have begun with

Dirichlet's introduction of Dirichlet *L*-functions to give the first proof of Dirichlet's theorem on arithmetic progressions. Another major milestone in the subject is the prime number theorem.

Analytic number theory can be split up into two major parts, divided more by the type of problems they attempt to solve than fundamental differences in technique. Multiplicative number theory deals with the distribution of the prime numbers, such as estimating the number of primes in an interval, and includes the prime number theorem and Dirichlet's theorem on primes in arithmetic progressions.

Additive number theory is concerned with the additive structure of the integers, such as Goldbach's conjecture that every even number greater than 2 is the sum of two primes. One of the main results in additive number theory is the solution to Waring's problem.

Developments within analytic number theory are often refinements of earlier techniques, which reduce the error terms and widen their applicability. For example, the *circle method* of Hardy and Littlewood was conceived as applying to power series near the unit circle in the complex plane; it is now thought of in terms of finite exponential sums (that is, on the unit circle, but with the power series truncated).

The needs of diophantine approximation are for auxiliary functions that aren't generating functions—their coefficients are constructed by use of a pigeonhole principle—and involve several complex variables. The fields of diophantine approximation and transcendence theory have expanded, to the point that the techniques have been applied to the Mordell conjecture.

The biggest technical change after 1950 has been the development of *sieve methods* as a tool, particularly in multiplicative problems. These are combinatorial in nature, and quite varied. The extremal branch of combinatorial theory has in return been greatly influenced by the value placed in analytic number theory on quantitative upper and lower bounds. Another recent development is *probabilistic number theory,* which uses tools from probability theory to estimate the distribution of number theoretic functions, such as how many prime divisors a number has.

Problems and Results in Analytic Number Theory

The great theorems and results within analytic number theory tend not to be exact structural results about the integers, for which algebraic and geometrical tools are more appropriate. Instead, they give approximate bounds and estimates for various number theoretical functions, as the following examples illustrate.

Multiplicative Number Theory

Euclid showed that there are an infinite number of primes but it is very difficult to find an efficient method for determining whether or not a number is prime, especially a large number. A related but easier problem is to determine the asymptotic distribution of the prime numbers; that is, a rough description of how many primes are smaller than a given number. Gauss, amongst others, after computing a large list of primes, conjectured that the number of primes less than or equal to a large number N is close to the value of the integral

$$\int_2^N \frac{1}{\log(t)} dt.$$

In 1859 Bernhard Riemann used complex analysis and a special meromorphic function now known as the Riemann zeta function to derive an analytic expression for the number of primes less than or equal to a real number x. Remarkably, the main term in Riemann's formula was exactly the above integral, lending substantial weight to Gauss's conjecture. Riemann found that the error terms in this expression, and hence the manner in which the primes are distributed, are closely related to the complex zeros of the zeta function. Using Riemann's ideas and by getting more information on the zeros of the zeta function, Jacques Hadamard and Charles Jean de la Vallée-Poussin managed to complete the proof of Gauss's conjecture. In particular, they proved that if

$$\pi(x) = (\text{number of primes } \leq x),$$

then

$$\lim_{x \to \infty} \frac{\pi(x)}{x / \log x} = 1.$$

This remarkable result is what is now known as the *Prime Number Theorem*. It is a central result in analytic number theory. Loosely speaking, it states that given a large number N, the number of primes less than or equal to N is about $N/\log(N)$.

More generally, the same question can be asked about the number of primes in any arithmetic progression $a+nq$ for any integer n. In one of the first applications of analytic techniques to number theory, Dirichlet proved that any arithmetic progression with a and q coprime contains infinitely many primes. The prime number theorem can be generalised to this problem; letting

$$\pi(x,a,q) = (\text{number of primes } \leq x \text{ such that } p \text{ is in the arithmetic progression } a+nq, n \in \mathbf{Z}),$$

then if a and q are coprime,

$$\lim_{x\to\infty} \frac{\pi(x,a,q)\phi(q)}{x/\log x} = 1.$$

There are also many deep and wide ranging conjectures in number theory whose proofs seem too difficult for current techniques, such as the Twin prime conjecture which asks whether there are infinitely many primes p such that $p + 2$ is prime. On the assumption of the Elliott-Halberstam conjecture it has been proven recently (by Daniel Goldston, János Pintz, Cem Yýldýrým) that there are infinitely many primes p such that $p + k$ is prime for some positive even k less than 16.

Additive Number Theory

One of the most important problems in additive number theory is Waring's problem, which asks whether it is possible, for any $k \geq 2$, to write any positive integer as the sum of a bounded number of kth powers,

$$n = x_1^k + \cdots + x_\ell^k.$$

The case for squares, $k = 2$, was answered by Lagrange in 1770, who proved that every positive integer is the sum of at most four squares. The general case was proved by Hilbert in 1909, using algebraic techniques which gave no explicit bounds. An important breakthrough was the application of analytic tools to the problem by Hardy and Littlewood. These techniques are known as the circle method, and give explicit upper bounds for the function $G(k)$, the smallest number of kth powers needed, such as Vinogradov's bound

$$G(k) \leq k(3 \log k + 11).$$

Diophantine Problems

Diophantine problems are concerned with integer solutions to polynomial equations: one may study the distribution of solutions, that is, counting solutions according to some measure of "size" or *height*.

An important example is the Gauss circle problem, which asks for integers points $(x\ y)$ which satisfy

$$x^2 + y^2 \leq r^2.$$

In geometrical terms, given a circle centred about the origin in the plane with radius r, the problem asks how many integer lattice points lie on or inside the circle. It is not hard to prove that the answer is $\pi r^2 + E(r)$, where $E(r)/r^2 \to 0$ as $r \to \infty$. Again, the difficult part and a great achievement of analytic number theory is obtaining specific upper bounds on the error term $E(r)$.

It was shown by Gauss that $E(r) = O(r)$. In general, an $O(r)$ error term would be possible with the unit circle (or, more properly, the closed unit disk) replaced by the dilates of any bounded planar region with piecewise smooth boundary. Furthermore, replacing the unit circle by the unit square, the error term for the general problem can be as large as a linear function of r.

Therefore getting an error bound of the form $O(r^\delta)$ for some $\delta < 1$ in the case of the circle is a significant improvement. The first to attain this was Sierpiñski in 1906, who showed $E(r) = O(r^{2/3})$. In 1915, Hardy and Landau each showed that one does *not* have $E(r) = O(r^{1/2})$. Since then the goal has been to show that for each fixed $\epsilon > 0$ there exists a real number $C(\epsilon)$ such that $E(r) \leq C(\epsilon) r^{1/2+\epsilon}$. In 2000 Huxley showed that $E(r) = O(r^{131/208})$, which is the best published result.

Methods of Analytic Number Theory

Dirichlet series: One of the most useful tools in multiplicative number theory are Dirichlet series, which are functions of a complex variable defined by an infinite series

$$f(s)=\sum_{n=1}^{\infty}a_n n^{-s}.$$

Depending on the choice of coefficients a_n, this series may converge everywhere, nowhere, or on some half plane. In many cases, even where the series does not converge everywhere, the holomorphic function it defines may be analytically continued to a meromorphic function on the entire complex plane. The utility of functions like this in multiplicative problems can be seen in the formal identity

$$\left(\sum_{n=1}^{\infty}a_n n^{-s}\right)\left(\sum_{n=1}^{\infty}b_n n^{-s}\right)=\sum_{n=1}^{\infty}\left(\sum_{k\ell=n}a_k b_\ell\right)n^{-s};$$

hence the coefficients of the product of two Dirichlet series are the multiplicative convolutions of the original coefficients. Furthermore, techniques such as partial summation and Tauberian theorems can be used to get information about the coefficients from analytic information about the Dirichlet series. Thus a common method for estimating a multiplicative function is to express it as a Dirichlet series (or a product of simpler Dirichlet series using convolution identities), examine this series as a complex function and then convert this analytic information back into information about the original function.

Riemann Zeta Function

Euler showed that the fundamental theorem of arithmetic implies that

$$\sum_{n=1}^{\infty}\frac{1}{n^s}=\prod_{p}^{\infty}\frac{1}{1-p^{-s}} \text{ for } s>1 \ (p \text{ is prime number})$$

Euler's proof of the infinity of prime numbers makes use of the divergence of the term at the left hand side for $s = 1$ (the so-called harmonic series), a purely analytic result. Euler was also the first to use analytical arguments for the purpose of studying properties of integers, specifically by constructing generating power series. This was the beginning of analytic number theory.

Later, Riemann considered this function for complex values of s and showed that this function can be extended to a meromorphic function on the entire plane with a simple pole at $s = 1$. This function is now known as the Riemann Zeta function and is denoted by $\zeta(s)$. There is a plethora of literature on this function and the function is a special case of the more general Dirichlet L-functions.

Analytic number theorists are often interested in the error of approximations such as the prime number theorem. In this case, the error is smaller than $x/\log x$. Riemann's formula for $\pi(x)$ shows that the error term in this approximation can be expressed in terms of the zeros of the zeta function. In his 1859 paper, Riemann conjectured that all the "non-trivial" zeros of ζ lie on the line $\Re(s) = 1/2$ but never provided a proof of this statement. This famous and long-standing conjecture is known as the *Riemann Hypothesis* and has many deep implications in number theory; in fact, many important theorems have been proved under the assumption that the hypothesis is true. For example under the assumption of the Riemann Hypothesis, the error term in the prime number theorem is $O(x^{1/2+\varepsilon})$.

In the early 20th century G.H.Hardy and Littlewood proved many results about the zeta function in an attempt to prove the Riemann Hypothesis. In fact, in 1914, Hardy proved that there were infinitely many zeros of the zeta function on the critical line

$$\Re(z) = 1/2.$$

This led to several theorems describing the density of the zeros on the critical line.

2

Algebraic Number Theory

Algebraic number theory is a major branch of number theory which studies algebraic structures related to algebraic integers. This is generally accomplished by considering a ring of algebraic integers O in an algebraic number field K/Q, and studying their algebraic properties such as factorization, the behaviour of ideals, and field extensions.

In this setting, the familiar features of the integers—such as unique factorization—need not hold. The virtue of the primary machinery employed—Galois theory, group cohomology, group representations, and L-functions—is that it allows one to deal with new phenomena and yet partially recover the behaviour of the usual integers.

Basic Notions

Unique factorization and the ideal class group: One of the first properties of Z that can fail in the ring of integers O of an algebraic number field K is that of the unique factorization of integers into prime numbers. The prime numbers in Z are generalized to irreducible elements in O, and though the unique factorization of elements of O into irreducible elements may hold in some cases (such as for the Gaussian integers Z[i]), it may also fail, as in the case of Z[√-5] where

$$6 = 2\cdot 3 = (1+\sqrt{-5})\cdot(1-\sqrt{-5}).$$

The ideal class group of O is a measure of how much unique factorization of elements fails; in particular, the ideal class group is trivial if, and only if, O is a unique factorization domain.

Factoring prime ideals in extensions

Unique factorization can be partially recovered for *O* in that it has the property of unique factorization of *ideals* into prime ideals (i.e. it is a Dedekind domain). This makes the study of the prime ideals in *O* particularly important. This is another area where things change from Z to *O*: the prime numbers, which generate prime ideals of Z (in fact, every single prime ideal of Z is of the form (p):=pZ for some prime number p,) may no longer generate prime ideals in *O*. For example, in the ring of Gaussian integers, the ideal 2Z[i] is no longer a prime ideal; in fact

$$2\mathbf{Z}[i] = \left((1+i)\mathbf{Z}[i]\right)^2.$$

On the other hand, the ideal 3Z[i] is a prime ideal. The complete answer for the Gaussian integers is obtained by using a theorem of Fermat's, with the result being that for an odd prime number p

$$p\mathbf{Z}[i] \text{ is a prime ideal if } p \equiv 3 (\text{mod} 4)$$

$$p\mathbf{Z}[i] \text{ is not a prime ideal if } p \equiv 1 (\text{mod} 4).$$

Generalizing this simple result to more general rings of integers is a basic problem in algebraic number theory. Class field theory accomplishes this goal when *K* is an abelian extension of Q (i.e. a Galois extension with abelian Galois group).

Primes and Places

An important generalization of the notion of prime ideal in *O* is obtained by passing from the so-called *ideal-theoretic* approach to the so-called *valuation-theoretic* approach. The relation between the two approaches arises as follows. In addition to the usual absolute value function $|\cdot| : Q \to R$, there are absolute value functions $|\cdot|_p : Q \to R$ defined for each prime number p in Z, called p-adic absolute values. Ostrowski's theorem states that these are all possible absolute value functions on Q (up to equivalence). This suggests that the usual absolute value could be considered as another prime.

More generally, a prime of an algebraic number field *K* (also called a place) is an equivalence class of absolute values on *K*. The primes in *K* are of two sorts: p-adic absolute values like $|\cdot|_p$, one for each prime ideal p of *O*, and absolute values like $|\cdot|$ obtained by considering K as a subset of the complex numbers in various possible ways and using the absolute value $|\cdot| : C \to R$. A prime of the first kind is called a finite prime (or finite place) and one of the second kind is called an infinite prime (or infinite place). Thus, the set of primes of Q is generally denoted { 2, 3, 5, 7, ..., " }, and the usual absolute value on Q is often denoted $|\cdot|_{"}$ in this context.

The set of infinite primes of K can be described explicitly in terms of the embeddings $K \to C$ (i.e. the non-zero ring homomorphisms from K to C). Specifically, the set of embeddings can be split up into two disjoint subsets, those whose image is contained in R, and the rest. To each embedding $\sigma : K \to R$, there corresponds a unique prime of K coming from the absolute value obtained by composing σ with the usual absolute value on R; a prime arising in this fashion is called a real prime (or real place). To an embedding $\tau : K \to C$ whose image is *not* contained in R, one can construct a distinct embedding τ, called the *conjugate embedding*, by composing τ with the complex conjugation map C $\to$ C. Given such a pair of embeddings τ and τ, there corresponds a unique prime of K again obtained by composing τ with the usual absolute value (composing τ instead gives the same absolute value function since $|z| = |z|$ for any complex number z, where z denotes the complex conjugate of z). Such a prime is called a complex prime (or complex place). The description of the set of infinite primes is then as follows: each infinite prime corresponds either to a unique embedding $\sigma : K \to R$, or a pair of conjugate embeddings $\tau, \tau : K \to C$. The number of real (respectively, complex) primes is often denoted r_1 (respectively, r_2). Then, the total number of embeddings $K \to C$ is r_1+2r_2 (which, in fact, equals the degree of the extension K/Q).

Units

The fundamental theorem of arithmetic describes the multiplicative structure of Z. It states that every non-zero integer can be written (essentially) uniquely as a product of prime powers and ±1. The unique factorization of ideals in the ring O recovers part of this description, but fails to address the factor ±1. The integers 1 and -1 are the invertible elements (i.e. units) of Z. More generally, the invertible elements in O form a group under multiplication called the unit group of O, denoted $O^{\times}$. This group can be much larger than the cyclic group of order 2 formed by the units of Z. Dirichlet's unit theorem describes the abstract structure of the unit group as an abelian group. A more precise statement giving the structure of $O^{\times} \otimes_{Z} Q$ as a Galois module for the Galois group of *K*/Q is also possible. The size of the unit group, and its lattice structure give important numerical information about *O*, as can be seen in the class number formula.

Local Fields

Completing a number field *K* at a place *w* gives a complete field. If the valuation is archimedean, one gets R or C, if it is non-archimedean and lies over a prime *p* of the rationals, one gets a finite extension K_w / Q_p: a complete, discrete valued field with finite residue field. This process simplifies the arithmetic of the field and allows the local study of problems. For example the Kronecker–Weber theorem can be deduced easily from the analogous local statement. The philosophy behind the study of local fields is largely motivated by geometric methods. In algebraic geometry, it is common to study varieties locally at a point by localizing to a maximal ideal. Global information can then be recovered by gluing together local data. This spirit is adopted in algebraic number theory. Given a prime in the ring of algebraic integers in a number field, it is desirable to study the field locally at that prime. Therefore one localizes the ring of algebraic integers to that prime and then completes the fraction field much in the spirit of geometry.

Major Results

Finiteness of the class group: One of the classical results in algebraic number theory is that the ideal class group of an algebraic number field *K* is finite. The order of the class group is called the class number, and is often denoted by the letter *h*.

Dirichlet's Unit Theorem

Dirichlet's unit theorem provides a description of the structure of the multiplicative group of units $O^{\times}$ of the ring of integers *O*. Specifically, it states that $O^{\times}$ is isomorphic to $G \times Z^r$, where *G* is the finite cyclic group consisting of all the roots of unity in *O*, and $r = r_1 + r_2 - 1$ (where r_1 (respectively, r_2) denotes the number of real embeddings (respectively, pairs of conjugate non-real embeddings) of K). In other words, $O^\times$ is a finitely generated abelian group of rank $r_1 + r_2 - 1$ whose torsion consists of the roots of unity in O.

Diophantine Geometry

In mathematics, diophantine geometry is one approach to the theory of Diophantine equations, formulating questions about such equations in terms of algebraic geometry over a ground field K that is not algebraically closed, such as the field of rational numbers or a finite field, or more general commutative ring such as the integers. A single equation defines a hypersurface, and simultaneous Diophantine equations give rise to a general algebraic variety V over K; the typical question is about the nature of the set $V(K)$ of points on V with co-ordinates in K, and by means of height functions quantitative questions about the "size" of these solutions may be posed, as well as the qualitative issues of whether any points exist, and if so whether there are an infinite number.

Given the geometric approach, the consideration of homogeneous equations and homogeneous co-ordinates is fundamental, for the same reasons that projective geometry is the dominant approach in algebraic geometry. Rational number solutions therefore are the primary consideration; but integral solutions (i.e. lattice points) can be treated in the same way as an affine variety may be considered inside a projective variety that has extra points at infinity.

The general approach of diophantine geometry is illustrated by Faltings' theorem (a conjecture of L. J. Mordell) starting that an algebraic curve C of genus $g > 1$ over the rational numbers has only finitely many rational points. The first result of this kind may have been the theorem of Hilbert and Hurwitz dealing with the case $g = 0$. The theory consists both of theorems and many conjectures and open questions.

Background

Serge Lang published a book *Diophantine Geometry* in the area, in 1962. The traditional arrangement of material on Diophantine equations was by degree and number of variables, as in Mordell's *Diophantine Equations* (1969). Mordell's book starts with a remark on homogeneous equations $f = 0$ over the rational field, attributed to C. F. Gauss, that non-zero solutions in integers (even primitive lattice points) exist if non-zero rational solutions do, and notes a caveat of L. E. Dickson, which is about parametric solutions.

The Hilbert-Hurwitz result from 1890 reducing the diophantine geometry of curves of genus 0 to degrees 1 and 2 (conic sections) occurs in Chapter 17, as does Mordell's conjecture. Siegel's theorem on integral points occurs in Chapter 28. Mordell's theorem on the finite generation of the group of rational points on an elliptic curve is in Chapter 16, and integer points on the Mordell curve in Chapter 26.

In a Hostile Review of Lang's Book, Mordell Urote

In recent times, powerful new geometric ideas and methods have been developed by means of which important new arithmetical theorems and related results have been found and proved and some of these are not easily proved otherwise. Further, there has been a tendency to clothe the old results, their extensions, and proofs in the new geometrical language.

Sometimes, however, the full implications of results are best described in a geometrical setting. Lang has these aspects very much in mind in this book, and seems to miss no opportunity for geometric presentation. This accounts for his title "Diophantine Geometry."

He notes that the content of the book is largely versions of the Mordell-Weil theorem, Thue-Siegel-Roth theorem, Siegel's theorem, with a treatment of Hilbert's irreducibility theorem and applications (in the style of Siegel). Leaving aside issues of generality, and a completely different style, the major mathematical difference between the two books is that Lang used abelian varieties and offered a proof of Siegel's theorem, while Mordell noted that the proof "is of a very advanced character" (p. 263).

Despite a bad press initially, Lang's conception has been sufficiently widely accepted for a 2006 tribute to call the book "visionary". A larger field sometimes called "arithmetic of algebraic varieties" now includes diophantine geometry with class field theory, complex multiplication, local zeta-functions and L-functions. Paul Vojta wrote:

While others at the time shared this viewpoint (e.g., Weil, Tate, Serre), it is easy to forget that others did not, as Mordell's review of *Diophantine Geometry* attests.

Recent Approaches and Subfields

The areas below date as such from no earlier than the mid-twentieth century, even if they are based on older material. For example, as is explained below, the matter of algorithms in number theory is very old, in some sense older than the concept of proof; at the same time, the modern study of computability dates only from the 1930s and 1940s, and computational complexity theory from the 1970s.

Probabilistic Number Theory

Probabilistic number theory is a subfield of number theory, which explicitly uses probability to answer questions of number theory. One basic idea underlying it is that different prime numbers are, in some serious sense, like independent random variables. This however is not an idea that has a unique useful formal expression.

The founders of the theory were Paul Erdõs, Aurel Wintner and Mark Kac during the 1930s, one of the most intense periods of investigation in analytic number theory. The Erdõs–Wintner theorem and the Erdõs–Kac theorem on additive functions were foundational results.

Arithmetic Combinatorics

Arithmetic combinatorics arose out of the interplay between number theory, combinatorics, ergodic theory and harmonic analysis. It is about combinatorial estimates associated with arithmetic operations (addition, subtraction, multiplication, and division). *Additive combinatorics* refers to the special case when only the operations of addition and subtraction are involved.

For example: if A is a set of N integers, how large or small can the sumset

$$A+A:=\{x+y:x,y\in A\},$$

the difference set

$$A-A:=\{x-y:x,y\in A\},$$

and the product set

$$A \times A := \{xy : x, y \in A\}$$

be, and how are the sizes of these sets related? (Not to be confused: the terms difference set and product set can have other meanings.)

The sets being studied may also belong to other spaces than the integers. e.g. groups, rings and fields.

Arithmetic combinatorics is explained in Green's review of "Additive Combinatorics" by Tao and Vu.

Additive Number Theory

In number theory, the specialty additive number theory studies subsets of integers and their behaviour under addition. More abstractly, the field of "additive number theory" includes the study of Abelian groups and commutative semigroups with an operation of addition. Additive number theory has close ties to combinatorial number theory and the geometry of numbers. Two principal objects of study are the sumset of two subsets $A$and $B$of elements from an Abelian group G,

$$A + B = \{a + b : a \in A, b \in B\},$$

and the h-fold sumset of A,

$$hA = \underbrace{A + \cdots + A}_{h}.$$

There are two main subdivisions listed below.

Number Theory

The first is principally devoted to consideration of direct problems over (typically) the integers, that is, to determining which elements can be represented as a summand from hA, where A is a fixed subset. Two classical problems of this type are the Goldbach conjecture (which is the conjecture that 2P contains all even numbers greater than two, where $P$is the set of primes) and Waring's problem (which asks how large must h be to guarantee that hA_k contains all positive integers, where

$$A_k = \{0^k, 1^k, 2^k, 3^k, \ldots\}$$

is the set of k-th powers). Many of these problems are studied using the tools from the Hardy-Littlewood circle method and from sieve methods. For example, Vinogradov proved that every sufficiently large odd number is the sum of three primes, and so every sufficiently large even integer is the sum of four primes. Hilbert proved that, for every integer $k > 1$, every nonnegative integer is the sum of a bounded number of k-th powers. In general, a set A of nonnegative integers is called a basis of order h if h Acontains all positive integers, and it is called an assymptotic basis if h Acontains all sufficiently large integers. Much current research in this area concerns properties of general asymptotic bases of finite order. For example, a set A is called a minimal asymptotic basis of order h if A is an asymptotic basis of order h but no proper subset of A is an asymptotic basis of order h. It has been proved that minimal asymptotic bases of order h exist for all h, and that there also exist asymptotic bases of order h that contain no minimal asymptotic bases of order h. Another question to be considered is how small can the number of representations of as a sum of elements in an asymptotic basis can be. This is the content of the Erdõs–Turán conjecture on additive bases.

Combinatorics

The second is principally devoted to consideration of inverse problems, often over more general groups than just the integers, that is, given some information about the sumset A+B, the aim is find information about the structure of the individual sets A and B. (A more recent name sometimes associated to this sub-division is additive combinatorics.) Unlike problems related to classical bases, as described above, this sub-area often deals with finite subsets rather than infinite ones. A typical question is what is the structure of a pair of subsets whose sumset has small cardinality (in relation to $|A|$ and $|B|$). In the case of the integers, the classical Freiman's theorem provides a potent partial answer to this question in terms of multi-dimensional arithmetic progressions. Another typical problem is simply to find a lower bound for $|A+B|$ in terms of $|A|$ and $|B|$ (this can be view as an inverse problem with the given information for being $A+B$ that is $|A+B|$ sufficiently small and the structural conclusion then being that either A or B is the empty set; such problems are often considered direct problems as well). Examples of this type include the Erdõs–Heilbronn Conjecture (for a restricted sumset) and the Cauchy-Davenport Theorem. The methods used for tackling such questions draw from across the spectrum of mathematics, including combinatorics, ergodic theory, analysis, graph theory, group theory, and linear algebraic and polynomial methods.

Computations in Number Theory

While the word *algorithm* goes back only to certain readers of al-Khwârizmî, careful descriptions of methods of solution are older than proofs: such methods (that is, algorithms) are as old as any recognisable mathematics—ancient Egyptian, Babylonian, Vedic, Chinese—whereas proofs appeared only with the Greeks of the classical period. An interesting early case is that of what we now call the Euclidean algorithm. In its basic form (namely, as an algorithm for computing the greatest common divisor) it appears as Proposition 2 of Book VII in *Elements*, together with a proof of correctness.

However, in the form that is often used in number theory (namely, as an algorithm for finding integer solutions to an equation $ax + by = c$, or, what is the same, for finding the quantities whose existence is assured by the Chinese remainder theorem) it first appears in the works of Âryabhama (5th–6th century CE) as an algorithm called *kumm„aka* ("pulveriser"), without a proof of correctness.

There are two main questions: "can we compute this?" and "can we compute it rapidly?". Anybody can test whether a number is prime or, if it is not, split it into prime factors; doing so rapidly is another matter. We now know fast algorithms for testing primality, but, in spite of much work (both theoretical and practical), no truly fast algorithm for factoring.

The difficulty of a computation can be useful: modern protocols for encrypting messages (e.g., RSA) depend on functions that are known to all, but whose inverses (a) are known only to a chosen few, and (b) would take one too long a time to figure out on one's own. For example, these functions can be such that their inverses can be computed only if certain large integers are factorized. While many difficult computational problems outside number theory are known, most working encryption protocols nowadays are based on the difficulty of a few number-theoretical problems. On a different note — some things may not be computable at all; in fact, this can be proven in some instances. For instance, in 1970, it was proven that there is no Turing machine which can solve all Diophantine equations.

There are thus some problems in number theory that will never be solved. We even know the shape of some of them, viz., Diophantine equations in nine variables; we simply do not know, and cannot know, which coefficients give us equations for which the following two statements are both true: there are no solutions, and we shall never know that there are no solutions.

Applications

The number-theorist Leonard Dickson (1874-1954) said "Thank God that number theory is unsullied by any application". Such a view is no longer applicable to number theory. In 1974 Donald Knuth said "...virtually every theorem in elementary number theory arises in a natural, motivated way in connection with the problem of making computers do high-speed numerical calculations". Elementary number theory is taught in discrete mathematics courses for computer scientists, and on the other hand number theory also has applications to the continuous in numerical analysis. As well as the well-known applications to cryptography, there are also applications to many other areas of mathematics.

Literature

Two of the most popular introductions to the subject are:

- G. H. Hardy; E. M. Wright (2008) [1938]. *An introduction to the theory of numbers* (rev. by D. R. Heath-Brown and J. H. Silverman, 6th ed.). Oxford University Press. ISBN 978-0-19-921986-5.
- Vinogradov, I. M. (2003) [1954]. *Elements of Number Theory* (reprint of the 1954 ed.). Mineola, NY: Dover Publications.

Hardy and Wright's book is a comprehensive classic, though its clarity sometimes suffers due to the authors' insistence on elementary methods. Vinogradov's main attraction consists in its set of problems, which quickly lead to Vinogradov's own research interests; the text itself is very basic and close to minimal.

Popular choices for a second textbook include:

- Borevich, A. I.; Shafarevich, Igor R. (1966). *Number theory*. Pure and Applied Mathematics. 20. Boston, MA: Academic Press. ISBN 978-0-12-117850-5. MR 0195803.
- Serre, Jean-Pierre (1996) [1973]. *A course in arithmetic*. Graduate texts in mathematics. 7. Springer. ISBN 978-0-387-90040-7.

Glossary of Arithmetic and Diophantine Geometry

This is a glossary of arithmetic and Diophantine geometry in mathematics, areas growing out of the traditional study of Diophantine equations to encompass large parts of number theory and algebraic geometry. Much of the theory is in the form of proposed conjectures, which can be related at various levels of generality.

Diophantine geometry in general is the study of algebraic varieties *V* over fields *K* that are finitely generated over their prime fields—including as of special interest number fields and finite fields—and over local fields. Of those, only the complex numbers are algebraically closed; over any other *K* the existence of points of *V* with co-ordinates in *K* is something to be proved and studied as an extra topic, even knowing the geometry of *V*.

Arithmetical or *arithmetic* (algebraic) geometry is a field with a less elementary definition. After the advent of scheme theory it could reasonably be defined as the study of Alexander Grothendieck's schemes *of finite type* over the spectrum of the ring of integers Z. This point of view has been very influential for around half a century; it has very widely been regarded as fulfilling Leopold Kronecker's ambition to have number theory operate only with rings that are quotients of polynomial rings over the integers (to use the current language of commutative algebra).

In fact scheme theory uses all sorts of auxiliary constructions that do not appear at all 'finitistic', so that there is little connection with 'constructivist' ideas as such. That scheme theory may not be the last word appears from continuing interest in the 'infinite primes' (the real and complex local fields), which do not come from prime ideals as the p-adic numbers do.

A

Abc Conjecture: The abc conjecture of Masser and Oesterlé attempts to state as much as possible about repeated prime factors in an equation $a + b = c$. For example 3 + 125 = 128 but the prime powers here are exceptional.

Arakelov Height: The *Arakelov height* on a projective space over the field of algebraic numbers is a global height function with local contributions coming from Fubini–Study metrics on the Archimedean fields an d the usual metric on the non-Archimedean fields.

Arakelov Theory: Arakelov theory is an approach to arithmetic geometry that explicitly includes the 'infinite primes'.

Artin L-functions: Artin L-functions are defined for quite general Galois representations. The introduction of étale cohomology in the 1960s meant that Hasse–Weil L-functions could be regarded as Artin L-functions for the Galois representations on l-adic cohomology groups.

B

Birch and Swinnerton-Dyer conjecture: The Birch and Swinnerton-Dyer conjecture on elliptic curves postulates a connection between the rank of an elliptic curve and the order of pole of its Hasse-Weil L-function. It has been an important landmark in Diophantine geometry since the mid-1960s, with important results such as the Coates–Wiles theorem, Gross–Zagier theorem and Kolyvagin's theorem.

Bombieri–Lang conjecture: Enrico Bombieri, Serge Lang and Paul Vojta have conjectured that algebraic varieties of general type do not have Zariski dense subsets of K-rational points, for K a finitely-generated field. This circle of ideas includes the understanding of *analytic hyperbolicity* and the Lang conjectures on that, and the Vojta conjectures. An *analytically holomorphic algebraic variety* V over the complex numbers is one such that no holomorphic mapping from the whole complex plane to it exists, that is not constant. Examples include compact Riemann surfaces of genus $g > 1$. Lang conjectured that V is analytically holomorphic if and only if all subvarieties are of general type.

C

Canonical Height: The canonical height on an abelian variety is a height function that is a distinguished quadratic form.

Chabauty's Method: Chabauty's method, based on p-adic analytic functions, is a special application but capable of proving cases of the Mordell conjecture for curves whose Jacobian's rank is less than its dimension. It developed ideas from Thoralf Skolem's method for an algebraic torus. (Other older methods for Diophantine problems include Runge's method.)

Coates–Wiles Theorem: The *Coates–Wiles theorem* states that an elliptic curve with complex multiplication by an imaginary quadratic field of class number 1 and positive rank has L-function with a zero at $s=1$. This is a special case of the Birch and Swinnerton-Dyer conjecture.

Crystalline Cohomology: Crystalline cohomology is a p-adic cohomology theory in characteristic p, introduced by Alexander Grothendieck to fill the gap left by étale cohomology which is deficient in using mod p coefficients in this case. It is one of a number of theories deriving in some way from Dwork's method (q.v.), and has applications outside purely arithmetical questions.

D

Diagonal Forms: Diagonal forms are some of the simplest projective varieties to study from an arithmetic point of view (including the Fermat varieties). Their local zeta-functions are computed in terms of Jacobi sums. Waring's problem is the most classical case.

Diophantine Dimension: The *Diophantine dimension* of a field is the smallest natural number k, if it exists, such that the field of is class C_k: that is, such that any homogeneous polynomial of degree d in N variables has a non-trivial zero whenever $N > d^k$. Algebraically closed fields are of Diophantine dimension 0; quasi-algebraically closed fields of dimension 1.

Discriminant of a Point: The *discriminant of a point* refers to two related concepts relative to a point P on an algebraic variety V defined over a number field K: the *geometric (logarithmic) discriminant* $d(P)$ and the *arithmetic discriminant*, defined by Vojta.

The difference between the two may be compared to the difference between the arithmetic genus of a singular curve and the geometric genus of the desingularisation. The arithmetic genus is larger than the geometric genus, and the height of a point may be bounded in terms of the arithmetic genus. Obtaining similar bounds involving the geometric genus would have significant consequences.

Dwork's Method: Bernard Dwork used distinctive methods of p-adic analysis, p-adic algebraic differential equations, Koszul complexes and other techniques that have not all been absorbed into general theories such as crystalline cohomology. He first proved the rationality of local zeta-functions, the initial advance in the direction of the Weil conjectures.

E

Étale Cohomology: The search for a Weil cohomology (q.v.) was at least partially fulfilled in the étale cohomology theory of Alexander Grothendieck and Michael Artin. It provided a proof of the functional equation for the local zeta-functions, and was basic in the formulation of the Tate conjecture (q.v.) and numerous other theories.

F

Fermat's Last Theorem: Fermat's last theorem, the most celebrated conjecture of Diophantine geometry, was proved by Andrew Wiles and Richard Taylor.

Flat Cohomology: Flat cohomology is, for the school of Grothendieck, one terminal point of development. It has the disadvantage of being quite hard to compute with. The reason that the flat topology has been considered the 'right' foundational topos for scheme theory goes back to the fact of faithfully-flat descent, the discovery of Grothendieck that the representable functors are sheaves for it (i.e. a very general gluing axiom holds).

Function field analogy: It was realised in the nineteenth century that the ring of integers of a number field has analogies with the affine coordinate ring of an algebraic curve or compact Riemann surface, with a point or more removed corresponding to the 'infinite places' of a number field. This idea is more precisely encoded in the theory that global fields should all be treated on the same basis. The idea goes further. Thus elliptic surfaces over the complex numbers, also, have some quite strict analogies with elliptic curves over number fields.

G

Geometric Class Field Theory: The extension of class field theory-style results on abelian coverings to varieties of dimension at least two is often called *geometric* class field theory.

Good reduction: Fundamental to local analysis in arithmetic problems is to *reduce modulo* all prime numbers p or, more generally, prime ideals. In the typical situation this presents little difficulty for almost all p; for example denominators of fractions are tricky, in that reduction modulo a prime in the denominator looks like division by zero, but that rules out only finitely many p per fraction. With a little extra sophistication, homogeneous coordinates allow clearing of denominators by multiplying by a common scalar. For a given, single point one can do this and not leave a common factor p.

However singularity theory enters: a non-singular point may become a singular point on reduction modulo p, because the Zariski tangent space can become larger when linear terms reduce to 0 (the geometric formulation shows it is not the fault of a single set of coordinates). *Good reduction* refers to the reduced variety having the same properties as the original, for example, an algebraic curve having the same genus, or a smooth variety remaining smooth. In general there will be a finite set S of primes for a given variety V, assumed smooth, such that there is otherwise a smooth reduced V_p over Z/pZ. For abelian varieties, good reduction is connected with ramification in the field of division points by the Néron–Ogg–Shafarevich criterion. The theory is subtle, in the sense that the freedom to change variables to try to improve matters is rather unobvious.

Grothendieck–Katz Conjecture: The Grothendieck–Katz p-curvature conjecture applies reduction modulo primes to algebraic differential equations, to derive information on algebraic function solutions. It is an open problem as of 2005. The initial result of this type was Eisenstein's theorem.

H

Hasse Principle: The Hasse principle states that solubility for a global field is the same as solubility in all relevant local fields. One of the main objectives of Diophantine geometry is to classify cases where the Hasse principle holds. Generally that is for a large number of variables, when the degree of an equation is held fixed. The Hasse principle is often associated with the success of the Hardy–Littlewood circle method. When the circle method works, it can provide extra, quantitative information such as asymptotic number of solutions. Reducing the number of variables makes the circle method harder; therefore failures of the Hasse principle, for example for cubic forms in small numbers of variables (and in particular for elliptic curves as cubic curves) are at a general level connected with the limitations of the analytic approach.

Hasse–Weil L-function: A Hasse–Weil L-function, sometimes called a *global* L-function, is an Euler product formed from local zeta-functions. The properties of such L-functions remain largely in the realm of conjecture, with the proof of the Taniyama–Shimura conjecture being a breakthrough. The Langlands philosophy is largely complementary to the theory of global L-functions.

Height Function: A height function in Diophantine geometry quantifies the size of solutions to Diophantine equations. Classical or naive height is defined in terms of ordinary absolute value on homogeneous coordinates: it is now usual to take a logarithmic scale, that is, height is proportional to the "algebraic complexity" or number of bits needed to store a point. Heights were initially developed by André Weil and D. G. Northcott. Innovations around 1960 were the Néron–Tate height and the realisation that heights were linked to projective representations in much the same way that ample line bundles are in pure geometry.

Hilbertian Fields: A Hilbertian field K is one for which the projective spaces over K are not thin sets in the sense of Jean-Pierre Serre. This is a geometric take on Hilbert's irreducibility theorem which shows the rational numbers are Hilbertian. Results are applied to the inverse Galois problem. Thin sets (the French word is *mince*) are in some sense analogous to the meagre sets (French *maigre*) of the Baire category theorem.

I

Igusa Zeta-function: An Igusa zeta-function, named for Jun-ichi Igusa, is a generating function counting numbers of points on an algebraic variety modulo high powers p^n of a fixed prime number p. General rationality theorems are now known, drawing on methods of mathematical logic.

Infinite Descent: Infinite descent was Pierre de Fermat's classical method for Diophantine equations. It became one half of the standard proof of the Mordell–Weil theorem, with the other being an argument with height functions (q.v.). Descent is something like division by two in a group of principal homogeneous spaces (often called 'descents', when written out by equations); in more modern terms in a Galois cohomology group which is to be proved finite.

Iwasawa Theory: Iwasawa theory builds up from the analytic number theory and Stickelberger's theorem as a theory of ideal class groups as Galois modules and p-adic L-functions (with roots in Kummer congruence on Bernoulli numbers). In its early days in the late 1960s it was called *Iwasawa's analogue of the Jacobian.*

The analogy was with the Jacobian variety J of a curve C over a finite field F (*qua* Picard variety), where the finite field has roots of unity added to make finite field extensions $F2$ The local zeta-function (q.v.) of C can be recovered from the points $J(F2\ )$ as Galois module.

In the same way, Iwasawa added p^n-power roots of unity for fixed p and with $n \to$ ", for his analogue, to a number field K, and considered the inverse limit of class groups, finding a p-adic L-function earlier introduced by Kubota and Leopoldt.

K

K-theory: Algebraic K-theory is on one hand a quite general theory with an abstract algebra flavour, and, on the other hand, implicated in some formulations of arithmetic conjectures.

L

Linear torus: A *linear torus* is a geometrically irreducible Zariski-closed subgroup of an affine torus (product of multiplicative groups).

Local zeta-function: A local zeta-function is a generating function for the number of points on an algebraic variety V over a finite field F, over the finite field extensions of F. According to the Weil conjectures (q.v.) these functions, for non-singular varieties, exhibit properties closely analogous to the Riemann zeta-function, including the Riemann hypothesis.

M

Mordell Conjecture: The Mordell conjecture is now the Faltings theorem, and states that a curve of genus at least two has only finitely many rational points. The Uniformity conjecture states that there should be a uniform bound on the number of such points, depending only on the genus and the field of definition.

Mordell–Lang Conjecture: The Mordell–Lang conjecture is a collection of conjectures of Serge Lang unifying the Mordell conjecture and Manin–Mumford conjecture in an abelian variety or semi-abelian variety.

Mordell–Weil theorem: The Mordell–Weil theorem is a foundational result stating that for an abelian variety A over a number field K the group $A(K)$ is a finitely-generated abelian group. This was proved initially for number fields K, but extends to all finitely-generated fields.

Modellic Variety: A Mordellic variety is an algebraic variety which has only finitely many points in any finitely generated field.

N

Naive Height: The *naive* or *classical height* of a vector of rational numbers is the maximum absoulte value of the vector of coprime integers obtained by multiplying through by a lowest common denominator. This may be used to define height on a point in projective space over Q, or of a polynomials, regarded as a vector of coefficients, or of an algebraic number, from the height of its minimal polynomial.

Néron Symbol: The *Néron symbol* is a bimultiplicative pairing between divisors and algebraic cycles on an Abelian variety used in Néron's formulation of the Néron–Tate height as a sum of local contributions.

Néron–Tate Height: The Néron–Tate height (also often referred to as the canonical height) on an abelian variety A is a height function (q.v.) that is essentially intrinsic, and an exact quadratic form, rather than approximately quadratic with respect to the addition on A as provided by the general theory of heights. It can be defined from a general height by a limiting process; there are also formulae, in the sense that it is a sum of local contributions.

O

Ordinary Reduction: An Abelian variety A of dimension d has *ordinary reduction* at a prime p if it has good reduction at p and in addition the p-torsion has rank d.

Q

Quasi-algebraic Closure: The topic of quasi-algebraic closure, i.e. solubility guaranteed by a number of variables polynomial in the degree of an equation, grew out of studies of the Brauer group and the Chevalley–Warning theorem. It stalled in the face of counterexamples.

R

Reduction *modulo* a prime number or ideal

S

Sato–Tate Conjecture: The Sato–Tate conjecture is a conjectural result on the distribution of Frobenius elements in the Tate modules of the elliptic curves over finite fields obtained from reducing a given elliptic curve over the rationals. It is due to Mikio Sato and John Tate (independently, and around 1960, published somewhat later) and is by now supported by very substantial evidence. It is a prototype for Galois representations in general.

Special Set: The *special set* in an algebraic variety is the subset in which one might expect to find many rational points. The precise definition varies according to context. One definition is the Zariski closure of the union of images of algebraic groups under non-trivial rational maps; alternatively one may take images of abelian varieities; another definition is the union of all subvarieties that are not of general type. For abelian varieties the definition would be the union of all translates of proper abelian subvarieties.

For a complex variety, the *holomorphic special set* is the Zariski closure of the images of all non-constant holomorphic maps from C. Lang conjectured that the analytic and algebraic special sets are equal.

Subspace Theorem: Schmidt's subspace theorem shows that points of small height in projective space lie in a finite number of hyperplanes. A quantitative form of the theorem, in which the number of subspaces containing all solutions, was also obtained by Schmidt, and the theorem was generalised by Schlickewei (1977) to allow more general absolute values on number fields. The theorem may be used to obtain results on Diophantine equations such as Siegel's theorem on integral points and solution of the S-unit equation.

T

Tamagawa Numbers: The direct Tamagawa number definition works well only for linear algebraic groups. There the Weil conjecture on Tamagawa numbers was eventually proved. For abelian varieties, and in particular the Birch–Swinnerton-Dyer conjecture (q.v.), the Tamagawa number approach to a local-global principle fails on a direct attempt, though it has had heuristic value over many years. Now a sophisticated equivariant Tamagawa number conjecture is a major research problem.

Tate Conjecture: The Tate conjecture (John Tate, 1963) provided an analogue to the Hodge conjecture, also on algebraic cycles, but well within arithmetic geometry. It also gave, for elliptic surfaces, an analogue of the Birch–Swinnerton-Dyer conjecture (q.v.), leading quickly to a clarification of the latter and a recognition of its importance.

Tate Curve: The Tate curve is a particular elliptic curve over the p-adic numbers introduced by John Tate to study bad reduction.

U

Uniformity Conjecture: The *unformity conjecture* states that for any number field K and $g > 2$, there is a uniform bound $B(g,K)$ on the number of K-rational points on any curve of genus g.

The conjecture would follow from the Bombieri–Lang conjecture.

V

Vojta Conjecture: The Vojta conjecture is a complex of conjectures by Paul Vojta, making analogies between Diophantine approximation and Nevanlinna theory.

W

Weights: The yoga of weights is a formulation by Alexander Grothendieck of analogies between Hodge theory and l-adic cohomology.

Weil Cohomology: The initial idea, later somewhat modified, for proving the Weil conjectures (q.v.), was to construct a cohomology theory applying to algebraic varieties over finite fields that would both be as good as singular homology at detecting topological structure, and have Frobenius mappings acting in such a way that the Lefschetz fixed-point theorem could be applied to the counting in local zeta-functions. For later history see motive (algebraic geometry), motivic cohomology.

Weil Conjectures: The Weil conjectures were three highly-influential conjectures of André Weil, made public around 1949, on local zeta-functions. The proof was completed in 1973. Those being proved, there remain extensions of the Chevalley–Warning theorem congruence, which comes from an elementary method, and improvements of Weil bounds, e.g. better estimates for curves of the number of points than come from Weil's basic theorem of 1940. The latter turn out to be of interest for Goppa codes.

Weil Distributions on Algebraic Varieties: André Weil proposed a theory in the 1920s and 1930s on prime ideal decomposition of algebraic numbers in co-ordinates of points on algebraic varieties. It has remained somewhat under-developed.

Weil Function: A *Weil function* on an algebraic variety is a real-valued function defined off some Cartier divisor which generalises the concept of Green's function in Arakelov theory. They are used in the construction of the local components of the Néron–Tate height.

Weil Height Machine: The *Weil height machine* is an effective procedure for assigning a height function to any divisor on smooth projective variety over a number field (or to Cartier divisors on non-smooth varieties).

3

Theorems in Number Theory

The 15 theorem of John H. Conway and W. A. Schneeberger (Conway–Schneeberger Fifteen Theorem), proved in 1993, states that if an integral quadratic form with integer matrix represents all positive integers up to 15, then it represents all positive integers. (All quadratic forms in this article are implicitly assumed to be positive definite.) The proof was complicated, and was never published. Manjul Bhargava found a much simpler proof which was published in 2000.

This result is the best possible, as there are such forms as, for example,

$$w^2 + 2x^2 + 5y^2 + 5z^2,$$

that represent all positive integers other than 15.

A quadratic form representing all positive integers is sometimes called universal. For example,

$$w^2 + x^2 + y^2 + z^2$$

is universal because every positive integer can be written as a sum of 4 squares, by Lagrange's four-square theorem. By the 15 theorem, to verify this it is sufficient to check that every positive integer up to 15 is a sum of 4 squares. (This does not give an alternative proof of Lagrange's theorem, because Lagrange's theorem is used in the proof of the 15 theorem.)

A more precise version of the 15 theorem says that if an integral quadratic form with integral matrix represents the numbers 1, 2, 3, 5, 6, 7, 10, 14, 15 (sequence A030050 in OEIS) then it represents all positive integers. Moreover, for each of these 9 numbers, there is such a quadratic form representing all positive integers except for this number.

There are two different notions of integrality for integral quadratic forms: it can be called integral if its associated symmetric matrix is integral ("integral matrix"), or if all its coefficients are integral ("integer valued"). For example, $x^2 + xy + y^2$ is integer valued but does not have integral matrix.

In 2005 Manjul Bhargava and Jonathan P. Hanke announced a proof of Conway's conjecture that a similar theorem holds for integer valued integral quadratic forms, with the constant 15 replaced by 290. The proof is forthcoming in Inventiones Mathematicae.

A more precise version states that if an integer valued integral quadratic form represents all the numbers 1, 2, 3, 5, 6, 7, 10, 13, 14, 15, 17, 19, 21, 22, 23, 26, 29, 30, 31, 34, 35, 37, 42, 58, 93, 110, 145, 203, 290 (sequence A030051 in OEIS) then it represents all positive integers, and for each of these 29 numbers there is such a quadratic form representing all positive integers with the exception of this one number.

Bhargava has found analogous criteria for an integral quadratic form to represent all primes (the set {2, 3, 5, 7, 11, 13, 17, 19, 23, 29, 31, 37, 41, 43, 47, 67, 73} (sequence A154363 in OEIS)) and for such a quadratic form to represent all positive odd integers (the set {1, 3, 5, 7, 11, 15, 33} (sequence A116582 in OEIS)).

Abouabdillah's Theorem

Abouabdillah's theorem refers to two distinct theorems in mathematics, proven by Moroccan mathematician Driss Abouabdillah: one in geometry and one in number theory.

Geometry

In geometry, similarities of an Euclidean space preserve circles and spheres. Conversely, Abouabdillah's theorem states that every injective or surjective transformation of a Euclidean space that preserves circles or spheres is a similarity.

More precisely:

Theorem. Let E be a Euclidean affine space of dimension at least 2. Then:

1. Every surjective mapping $f: E \to E$ that transforms any four concyclic points into four concyclic points is a similarity.

2. Every injective mapping $f: E \to E$ that transforms any circle into a circle is a similarity.

Number Theory

The number-theoretic theorem of Abouabdillah is about antichains in the partially ordered set E_N consisting of the positive integers in the interval $[1,N]$, partially ordered by divisibility. With this partial order, an antichain is a set of integers within this interval, such that no member of this set is a divisor of any other member. It possible to prove using ideas related to Dilworth's theorem that the maximum number of elements in an antichain of E_{2n} is exactly n: there exists an antichain of this size consisting of all the numbers in the subinterval $[n + 1,2n]$, so the maximum size of an antichain is at least n.

However, there are only n odd numbers within the interval $[1,2n]$, for each odd number c in this interval at most one number of the form $2^k c$ may belong to any antichain, and every number in the interval has this form for some c, so the maximum size of an antichain is also at most n.

Abouabdillah's theorem characterizes more precisely the numbers that may belong to an antichain of maximum size in E_{2n}. Specifically, if x is any integer in the interval $[1,2n]$, decompose x as the product of a power of two and an odd number: $x = 2^k c$, where c is odd. Then, according to Abouabdillah's theorem, there exists an antichain of cardinality n in E_{2n} that contains x if and only if $2n < 3^{k+1}c$.

The smallest value in any maximum antichain of E_{2n} is at least 2^k, where 3^{k+1} is the first power of three that is greater than $2n$, as had been posed as a problem by Paul Erdõs (1937) and solved by Emma Lehmer (1939). Lehmer's solution immediately implies the special case of Abouabdillah's theorem for $c = 1$. Abouabdillah's theorem generalizes this solution to all values within the given interval.

Apéry's Theorem

In mathematics, Apéry's theorem is a result in number theory that states the Apéry's constant $\zeta(3)$ is irrational. That is, the number

$$\zeta(3)=\sum_{n=1}^{\infty}\frac{1}{n^3}=\frac{1}{1^3}+\frac{1}{2^3}+\frac{1}{3^3}+\ldots=1.2020569\ldots$$

cannot be written as a fraction p/q with p and q being integers.

History

Euler proved in the eighteenth century that if n is a positive integer then we have

$$\frac{1}{1^{2n}}+\frac{1}{2^{2n}}+\frac{1}{3^{2n}}+\frac{1}{4^{2n}}+\ldots=\frac{p}{q}\pi^{2n}$$

for some rational number p/q. Specifically, writing the infinite series on the left as $\zeta(2n)$ he showed

$$\zeta(2n)=(-1)^{n+1}\frac{B_{2n}(2\pi)^{2n}}{2(2n)!}$$

where the B_n are the rational Bernoulli numbers. Once it was proved that π^n is always irrational this showed that $\zeta(2n)$ is irrational for all positive integers n.

No such representation in terms of π is known for the so-called odd zeta constants, the values $\zeta(2n+1)$ for positive integers n. It has been conjectured that the ratios of these quantities

$$\frac{\zeta(2n+1)}{\pi^{2n+1}},$$

are transcendental for every integer n e ≥ 1.

Because of this, no proof could be found to show that the odd zeta constants were irrational, even though they were—and still are—all believed to be transcendental. However, in June 1978 Roger Apéry gave a talk entitled "Sur l'irrationalité de $\zeta(3)$." During the course of the talk he outlined proofs that $\zeta(3)$ and $\zeta(2)$ were irrational, the latter using methods simplified from those used to tackle the former rather than relying on the expression in terms of π. Due to the wholly unexpected nature of the result and

Apéry's blasé and very sketchy approach to the subject many of the mathematicians in the audience dismissed the proof as flawed. Three of the audience members suspected Apéry was onto something, though, and set out to confirm his proof.

Two months later these three—Henri Cohen, Hendrik Lenstra, and Alfred van der Poorten—finished their work, and on August 18 Cohen delivered a lecture giving full details of Apéry's proof. Following the talk Apéry himself took to the podium to explain the source of some of his ideas.

Apéry's Proof

Apéry's original proof was based on the well known irrationality criterion from Dirichlet, which states that a number ξ is irrational if there are infinitely many coprime integers p and q such that

$$\left|\xi - \frac{p}{q}\right| < \frac{c}{q^{1+\delta}}$$

for some fixed $c,\delta>0$.

The starting point for Apéry was the series representation of $\zeta(3)$ as

$$\zeta(3) = \frac{5}{2}\sum_{n=1}^{\infty}\frac{(-1)^{n-1}}{n^3\binom{2n}{n}}.$$

Roughly speaking, Apéry then defined a sequence $c_{n,k}$ which converges to $\zeta(3)$ about as fast as the above series, specifically

$$c_{n,k} = \sum_{m=1}^{n}\frac{1}{m^3} + \sum_{m=1}^{k}\frac{(-1)^{m-1}}{2m^3\binom{n}{m}\binom{n+m}{m}}.$$

He then defined two more sequences a_n and b_n that, roughly, have the quotient $c_{n,k}$. These sequences were

$$a_n = \sum_{k=0}^{n} c_{n,k}\binom{n}{k}^2\binom{n+k}{k}^2$$

and

$$b_n = \sum_{k=0}^{n} \binom{n}{k}^2 \binom{n+k}{k}^2.$$

The sequence a_n/b_n converges to $\zeta(3)$ fast enough to apply the criterion, but unfortunately a_n is not an integer after n=2. Nevertheless, Apéry showed that even after multiplying a_n and b_n by a suitable integer to cure this problem the convergence was still fast enough to guarantee irrationality.

Later Proofs

Within a year of Apéry's result an alternative proof was found by Frits Beukers, who replaced Apéry's series with integrals involving the shifted Legendre polynomials $\widetilde{P}_n(x)$. Using a representation that would later be generalized to Hadjicostas's formula, Beukers showed that

$$\int_0^1 \int_0^1 \frac{-\log(xy)}{1-xy} \widetilde{P}_n(x) \widetilde{P}_n(y) dx dy = \frac{A_n + B_n \zeta(3)}{\operatorname{lcm}[1, \ldots, n]^3}$$

for some integers A_n and B_n (sequences *OEIS* A171484 and *OEIS* A171485). Using partial integration and the assumption that $\zeta(3)$ was rational and equal to a/b, Beukers eventually derived the inequality

$$0 < \frac{1}{b} \le \left| A_n + B_n \zeta(3) \right| \le 4 \left(\frac{4}{5} \right)^n,$$

which is a contradiction since the right-most expression tends to zero and so must eventually fall below $1/b$.

A more recent proof by Wadim Zudilin is more reminiscent of Apéry's original proof, and also has similarities to a fourth proof by Yuri Nesterenko. These later proofs again derive a contradiction from the assumption that $\zeta(3)$ is rational by constructing sequences that tend to zero but are bounded below by some positive constant. They are somewhat less transparent than the earlier proofs, relying as they do on hypergeometric series.

Higher Zeta Constants

Apéry and Beukers could simplify their proofs to work on $\zeta(2)$ as well thanks to the series representation

$$\zeta(2) = 3\sum_{n=1}^{\infty} \frac{1}{n^2 \binom{2n}{n}}.$$

Due to the success of Apéry's method a search was undertaken for a number ξ_5 with the property that

$$\zeta(5) = \xi_5 \sum_{n=1}^{\infty} \frac{(-1)^{n-1}}{n^5 \binom{2n}{n}}.$$

If such a ξ_5 were found then the methods used to prove Apéry's theorem would be expected to work on a proof that $\zeta(5)$ is irrational. Unfortunately, extensive computer searching has failed to find such a constant, and in fact it is now known that if ξ_5 exists and if it is an algebraic number of degree at most 25, then the coefficients in its minimal polynomial must be enormous, at least 10, so extending Apéry's proof to work on the higher odd zeta constants doesn't seem likely to work.

Despite this, many mathematicians working in this area expect a breakthrough sometime soon. Indeed, recent work by Wadim Zudilin and Tanguy Rivoal has shown that infinitely many of the numbers $\zeta(2n+1)$ must be irrational, and even that at least one of the numbers $\zeta(5)$, $\zeta(7)$, $\zeta(9)$, and $\zeta(11)$ must be irrational. Their work uses linear forms in values of the zeta function and estimates upon them to bound the dimension of a vector space spanned by values of the zeta function at odd integers. Hopes that Zudilin could cut his list further to just one number did not materialise, but work on this problem is still an active area of research. Higher zeta constants have application to physics: they describe correlation functions in quantum spin chains.

Ax–Kochen Theorem

The Ax–Kochen theorem, named for James Ax and Simon B. Kochen, states that for each positive integer d there is a finite set

Y_d of prime numbers, such that if p is any prime not in Y_d then every homogeneous polynomial of degree d over the p-adic numbers in at least d^2+1 variables has a nontrivial zero.

The Proof of the Theorem

The proof of the theorem makes extensive use of methods from mathematical logic, such as model theory. One first proves Serge Lang's theorem, stating that the analogous theorem is true for the field $F_p((t))$ of formal Laurent series over a finite field F_p with $Y_d = \varnothing$. In other words, every homogeneous polynomial of degree d with more than d^2 variables has a non-trivial zero (so $F_p((t))$ is a C_2 field).

Then one shows that if two Henselian valued fields have equivalent valuation groups and residue fields, and the residue fields have characteristic 0, then they are elementarily equivalent (which means that a first order sentence is true for one if and only if it is true for the other).

Next one applies this to two fields, one given by an ultraproduct over all primes of the fields $F_p((t))$ and the other given by an ultraproduct over all primes of the p-adic fields Q_p. Both residue fields are given by an ultraproduct over the fields F_p, so are isomorphic and have characteristic 0, and both value groups are the same, so the ultraproducts are elementarily equivalent. (Taking ultraproducts is used to force the residue field to have characteristic 0; the residue fields of $F_p((t))$ and Q_p both have non-zero characteristic p.)

The elementary equivalence of these ultraproducts implies that for any sentence in the language of valued fields, there is a finite set Y of exceptional primes, such that for any p not in this set the sentence is true for $F_p((t))$ if and only if it is true for the field of p-adic numbers. Applying this to the sentence stating that every non-constant homogeneous polynomial of degree d in at least d^2+1 variables represents 0, and using Lang's theorem, one gets the Ax-Kochen theorem.

Alternative Proof

In 2008, Jan Denef found a purely geometric proof for a conjecture of Jean-Louis Colliot-Thélène which generalizes the

Ax-Kochen theorem. He presented his proof at the "Variétés rationnelles" seminar at École Normale Supérieure in Paris, but the proof has not been published yet.

Exceptional Primes

Emil Artin conjectured this theorem without the finite exceptional set Y_d, but Guy Terjanian found the following 2-adic counterexample for $d = 4$. Define

$$G(x) = G(x_1, x_2, x_3) = \Sigma\, x_i^4 - \Sigma_{i<j}\, x_i^2 x_j^2 - x_1 x_2 x_3 (x_1 + x_2 + x_3).$$

Then G has the property that it is 1 mod 4 if some x is odd, and 0 mod 16 otherwise. It follows easily from this that the homogeneous form

$$G(x) + G(y) + G(z) + 4G(u) + 4G(v) + 4G(w)$$

of degree d=4 in 18> d^2 variables has no non-trivial zeros over the 2-adic integers.

Later Terjanian showed that for each prime p and multiple d>2 of $p(p-1)$, there is a form over the p-adic numbers of degree d with more than d^2 variables but no nontrivial zeros. In other words, for all d> 2, Y_d contains all primes p such that $p(p-1)$ divides d.

Brown (1978) gave an explicit but very large bound for the exceptional set of primes p. If the degree d is 1, 2, or 3 the exceptional set is empty. Heath-Brown (2010) showed that if d=5 the exceptional set is bounded by 13, and Wooley (2008) showed that for d=7 the exceptional set is bounded by 883 and for d=11 it is bounded by 8053.

Baker's Theorem

In transcendence theory, a mathematical discipline, Baker's theorem gives a lower bound for linear combinations of logarithms of algebraic numbers.

The result, proved by Alan Baker (1966, 1967a, 1967b), subsumed many earlier results in transcendental number theory and solved a problem posed by Alexander Gelfond nearly fifteen years earlier. Baker used this to prove the transcendence of many numbers, and to derive effective bounds for the solutions of some Diophantine equations, and to solve the problem of finding all imaginary quadratic fields with class number 1.

History

To simplify notation we introduce the set L of logarithms of nonzero algebraic numbers, that is

$$L = \{\lambda \in \mathbb{C}: e^{\lambda} \in \overline{\mathbb{Q}}^{\times}\}.$$

Using this notation several results in transcendental number theory become much easier to state, for example the Hermite–Lindemann theorem becomes the statement that any nonzero element of L is transcendental.

In 1934, Alexander Gelfond and Theodor Schneider independently proved the Gelfond–Schneider theorem. This result is usually stated as: if a is algebraic and not equal to 0 or 1, and if b is algebraic and irrational, then a^b is transcendental. Equivalently, though, it says that if λ_1 and λ_2 are elements of L that are linearly independent over the rational numbers, then they are linearly independent over the algebraic numbers. So if λ_1 and λ_2 are elements of L and λ_2 isn't zero, then the quotient λ_1/λ_2 is either a rational number or transcendental. It can't be an algebraic irrational number like "2.

Although proving this result of "rational linear independence implies algebraic linear independence" for two elements of L was sufficient for his and Schneider's result, Gelfond felt that it was crucial to extend this result to arbitrarily many elements of L. Indeed, from Gelfond (1952, p. 177):

...one may assume ... that the most pressing problem in the theory of transcendental numbers is the investigation of the measures of transcendence of finite sets of logarithms of algebraic numbers.

This problem was solved fourteen years later by Alan Baker and has since had numerous applications not only to transcendence theory but in algebraic number theory and the study of Diophantine equations as well. Baker received the Fields medal in 1970 for both this work and his applications of it to Diophantine equations.

Statement

With the above notation, Baker's theorem is a nonhomogeneous generalisation of the Gelfond–Schneider theorem. Specifically it states:

If $\lambda_1,\ldots,\lambda_n$ are elements of L that are linearly independent over the rational numbers, then for any algebraic numbers $\beta_0,\ldots,\beta_n$, not all zero, we have

$$|\beta_0+\beta_1\lambda_1+\cdots+\beta_n\lambda_n|>H^{-C}$$

where H is the maximum of the heights of the β's and C is an effectively computable number depending on n, the λ's, and the maximum d of the degrees of the β's. (If β_0 is nonzero then the assumption that the λ's are linearly independent can be dropped.) In particular this number is nonzero, so 1 and the λ's are linearly independent over the algebraic numbers.

Just as the Gelfond–Schneider theorem is equivalent to the statement about the transcendence of numbers of the form a^b, so too Baker's theorem implies the transcendence of numbers of the form

$$a_1^{b_1}\cdots a_n^{b_n},$$

where the b_i are all algebraic, irrational, and 1, $b_1,\ldots,b_n$ are linearly independent over the rationals, and the a_i are all algebraic and not 0 or 1.

Baker (1977) also gave several versions with explicit constants. For example, if $e^{\lambda_j}=\alpha_j$ has height at most $A_j e-4$ and all the numbers β_j have height at most $Be-4$ then the linear form

$$\Lambda=\beta_0+\beta_1\lambda_1+\cdots\beta_n\lambda_n$$

is either 0 or satisfies

$$\log|\Lambda|>(16nd)^{200n}\Omega(\log\Omega-\log\log A_n)(\log B+\log\Omega)$$

where

$$\Omega=\log A_1\log A_2\ldots\log A_n$$

and the field generated by all the α's and β's over the rationals has degree at most d. In the special case when $\beta_0=0$ and all the β_j are rational integers, the rightmost term log Ω can be deleted.

An explicit result by Baker and Wüstholz for a linear form ∧with integer coefficients yields a lower bound of the form

$$\log|\Lambda|>-C\cdot h(\alpha_1)h(\alpha_2)\cdots h(\alpha_n)\log\max\{|\beta_1|,\ldots,|\beta_n|\},$$

with a constant C

$$C = 18(n+1)! \cdot n^{n+1} \cdot (32d)^{n+2} \log(2nd),$$

where $d$is the degree of the number field generated by the α_i.

Baker's Method

Baker's proof of his theorem is an extension of the argument given by Gel′fond. The main ideas of the proof are illustrated by the proof of the following qualitative version of the theorem of Baker (1966) described by Serre (1971): if the numbers $2\pi i$ and log $a_1,\ldots,$ log a_n are linearly independent over the rational numbers, for nonzero algebraic numbers $a_1,\ldots, a_n$, then they are linearly independent over the algebraic numbers. The precise quantitative version of Bakers theory can be proved by replacing the conditions that things are zero by conditions that things are sufficiently small throughout the proof.

The main idea of Bakers proof is to construct an auxiliary function $\Phi(z_1,\ldots,z_{n-1})$ of several variables that vanishes to high order at many points of the form $\Phi(l,l,\ldots,l)$, then repeatedly show that it vanishes to lower order at even more points of this form. Finally the fact that it vanishes (to order 1) at enough points of this form implies using Vandermonde determinants that there is a multiplicative relation between the numbers a_i.

Construction of the Function Φ

Assume there is a relation

$$\beta_1 \log \alpha_1 + \cdots + \beta_{n-1} \log \alpha_{n-1} = \log \alpha_n$$

for algebraic numbers $\alpha_1,\ldots,\alpha_n$, $\beta_1,\ldots,\beta_{n-1}$ The function Φ is of the form

$$\Phi(z_1,\ldots,z_{n-1}) = \sum_{\lambda_1=0}^{L} \cdots \sum_{\lambda_n=0}^{L} p(\lambda_1 \ldots, \lambda_n) \alpha_1^{(\lambda_1+\lambda_n\beta_1)z_1} \cdots \alpha_{n-1}^{(\lambda_{n-1}+\lambda_n\beta_{n-1})z_{n-1}}$$

The integer coefficients p are chosen so that they are not all zero and Φ and its derivatives of order at most some constant M vanish at $z_1=\ldots=z_{n-1} = l$, for integers l with $0 \le l \le h$ for some constant h. This is possible because these conditions are homogeneous linear equations in the coefficients p, which have a non-zero solution provided the number of unknown variables p is larger than the number of equations. The linear relation between the logs of the α's is needed to cut down the number of linear equations that have to be satisfied. Moreover, using Siegel's lemma, the sizes of the coefficients p can be chosen to be not too large. The constants L, h, and M have to be carefully adjusted to that the next part of the proof works, and are subject to some constraints, which are roughly:

- L must be somewhat smaller than M to make the argument about extra zeros below work.
- A small power of h must be larger than L to make the final step of the proof work.
- L^n must be larger than about $M^{n-1}h$ in order that it is possible to solve for the coefficients p.

The constraints can be satisfied by taking h to be sufficiently large, M to be some fixed power of h, and L to be a slightly smaller power of h. Baker took M to be about h^2 and L to be about $h^{2''1/2n}$.

The linear relation between the logarithms of the α's is used to reduce L slightly; roughly speaking, without it the condition L^n must be larger than about $M^{n-1}h$ would become L^n must be larger than about $M^n h$, which is incompatible with the condition that L is somewhat smaller than M.

Zeros of $\Phi(l,l,...l)$

The next step is to show that Φ vanishes to slightly smaller order at many more points of the form $z_1 = ...z_{n-1} = l$ for integers l. This idea was Baker's key innovation: previous work on this problem involved trying to increase the number of derivatives that vanish while keeping the number of points fixed, which does not seem to work in the multivariable case. This is done by combining two ideas; First one shows that the derivatives at these points are quite small, by using the fact that many derivatives of Φ vanish at many nearby points. Then one shows that derivatives of Φ at this point are given by algebraic integers times known constants. If an algebraic integer has all its conjugates bounded by a known constant, then it cannot be too small unless it is zero, because the product of all conjugates of a nonzero algebraic integer is at least 1 in absolute value. Combining these two ideas implies that Φ

vanishes to slightly smaller order at many more points $z_1 = \ldots z_{n-1} = l$. This part of the argument requires that Φ does not increase too rapidly; the growth of Φ depends on the size of *L*, so requires a bound on the size of *L*, which turns out to be roughly that *L* must be somewhat smaller than *M*.

More precisely, Baker showed that since Φ vanishes to order *M* at *h* consecutive integers, it also vanishes to order *M*/2 at $h^{1+1/8n}$ consecutive integers 1, 2, 3,.... Repeating this argument *J* times shows that Φ vanishes to order $M/2^J$ at $h^{1+J/8n}$ points, provided that *h* is sufficiently large and *L* is somewhat smaller than $M/2^J$.

One then takes *J* large enough that $h^{1+J/8n} > (L+1)^n$ (*J* larger than about 16*n* will do if $h^2 > L$) so that that $\Phi(l,...l)=0$ for all integers *l* with $1 \le l \le (L+1)^n$.

Completion of the Proof

The condition that $\Phi(l,...l)=0$ for all integers *l* with $1 \le l \le (L+1)^n$ can be written as

$$\sum_{\lambda_1=0}^{L} \cdots \sum_{\lambda_n=0}^{L} p(\lambda_1 \ldots, \lambda_n)\alpha_1^{\lambda_1 l} \cdots \alpha_n^{\lambda_n l} = 0$$

This consists of $(L+1)^n$ homogeneous linear equations in the $(L+1)^n$ unknowns *p*, and by assumption has a non-zero solution *p*, so the determinant of the matrix of coefficients must vanish. However this matrix is a Vandermonde matrix, so the formula for the determinant of such a matrix forces an equality two of the values

$$\alpha_1^{\lambda_1} \cdots \alpha_n^{\lambda_n}$$

so the numbers $\alpha_1,...,\alpha_n$ are multiplicatively dependent. Taking logs then shows that $2\pi i$, $\log \alpha_1,...,\log \alpha_n$ are linearly dependent over the rationals.

Extensions and Generalizations

Baker (1966) in fact gave a quantitative version of the theorem, giving effective lower bounds for the linear form in logarithms. This is done by a similar argument, except statements about something being zero are replaced by statements giving a small upper bound for it, and so on.

Baker (1967a) showed how to eliminate the assumption about $2\pi i$ in the theorem. This requires a modification of the final step of the proof. One shows that many derivatives of the function $\varphi(z)=\Phi(z,...z)$ vanish at $z=0$, by an argument similar to the one above. But these equations for the first $(L+1)^n$ derivatives again give a homogeneous set of linear equations for the coefficients p, so the determinant is zero, and is again a Vandermonde determinant, this time for the numbers $\lambda_1\log\alpha_1+...\lambda_n\log\alpha_n$. So two of these expressions must be the same which shows that $\log\alpha_1,...,\log\alpha_n$ are linearly dependent over the rationals.

Baker (1967b) gave an inhomogeneous version of the theorem, showing that $\beta_0+\beta_1\log\alpha_1+...+\beta_n\log\alpha_n$ is nonzero for nonzero algebraic numbers $\beta_0, ..., \beta_n, \alpha_1,..., \alpha_n$, and moreover giving an effective lower bound for it. The proof is similar to the homogeneous case: one can assume that

$$\beta_0+\beta_1\log\alpha_1+\cdots+\beta_{n-1}\log\alpha_{n-1}=\log\alpha_n$$

and one inserts an extra variable z_0 into Φ as follows:

$$\Phi(z_0,\ldots,z_{n-1})=\sum_{\lambda_0=0}^{L}\cdots\sum_{\lambda_n=0}^{L}p(\lambda_0\ldots,\lambda_n)$$

$$z_0^{\lambda_0}e^{\lambda_n\beta_0z_0}\alpha_1^{(\lambda_1+\lambda_n\beta_1)z_1}\cdots\alpha_{n-1}^{(\lambda_{n-1}+\lambda_n\beta_{n-1})z_{n-1}}$$

Corollaries

As mentioned above, the theorem includes numerous earlier transcendence results concerning the exponential function, such as the Hermite–Lindemann theorem and Gelfond–Schneider theorem. It is not quite as encompassing as the still unproven Schanuel's conjecture, and does not imply the six exponentials theorem nor, clearly, the still open four exponentials conjecture.

The main reason Gelfond desired an extension of his result was not just for a slew of new transcendental numbers. In 1935 he used the tools he had developed to prove the Gelfond–Schneider theorem to derive a lower bound for the quantity

$$|\beta_1\lambda_1+\beta_2\lambda_2|$$

where β_1 and β_2 are algebraic and λ_1 and λ_2 are in L. Baker's proof gave lower bounds for quantities like the above but with arbitrarily many terms, and he could use these bounds to develop effective means of tackling Diophantine equations and to solve Gauss' class number problem.

Extensions

Baker's theorem grants us the linear independence over the algebraic numbers of logarithms of algebraic numbers. This is weaker than proving their *algebraic* independence. So far no progress has been made on this problem at all. It has been conjectured that if $\lambda_1,\ldots,\lambda_n$ are elements of L that are linearly independent over the rational numbers, then they are algebraically independent too. This is a special case of Schanuel's conjecture, but so far it remains to be proved that there even exist two algebraic numbers whose logarithms are algebraically independent. Indeed, Baker's theorem rules out linear relations between logarithms of algebraic numbers unless there are trivial reasons for them; the next most simple case, that of ruling out homogeneous quadratic relations, is the still open four exponentials conjecture.

Similarly, extending the result to algebraic independence but in the p-adic setting, and using the p-adic logarithm function, remains an open problem. It is known that proving algebraic independence of linearly independent p-adic logarithms of algebraic p-adic numbers would prove Leopoldt's conjecture on the p-adic ranks of units of a number field.

4

Basic Theorems of Number

Beatty Sequence

In mathematics, a Beatty sequence (or homogeneous Beatty sequence) is the sequence of integers found by taking the floor of the positive multiples of a positive irrational number. Beatty sequences are named after Samuel Beatty, who wrote about them in 1926.

Rayleigh's theorem, named after Lord Rayleigh, states that the complement of a Beatty sequence, consisting of the positive integers that are not in the sequence, is itself a Beatty sequence generated by a different irrational number.

Beatty sequences can also be used to generate Sturmian words.

Definition

A positive irrational number r generates the Beatty sequence

$$\mathcal{B}_r = \lfloor r \rfloor, \lfloor 2r \rfloor, \lfloor 3r \rfloor, \ldots = (\lfloor nr \rfloor)_{n\geq 1}$$

If $r > 1$, then $s = r/(r-1)$ is also a positive irrational number. They naturally satisfy

$$\frac{1}{r} + \frac{1}{s} = 1$$

and the sequences

$$\mathcal{B}_r = (\lfloor nr \rfloor)_{n\geq 1} \text{ and}$$

$$\mathcal{B}_s = (\lfloor ns \rfloor)_{n\geq 1}$$

form a *pair of complementary Beatty sequences.*

A more general non-homogeneous Beatty sequence takes the form

$$\mathcal{B}_r = \lfloor r + p \rfloor, \lfloor 2r + p \rfloor, \lfloor 3r + p \rfloor, \ldots = (\lfloor nr + p \rfloor)_{n \geq 1}$$

where *p*is a real number. For $p = 1$, the complementary non-homogeneous Beatty sequences can be found by making $t = 1/r$ so that

$$\mathcal{B}_r = (\lfloor n(r+1) \rfloor)_{n \geq 1} \text{ and}$$

$$\mathcal{B}_t = (\lfloor n(t+1) \rfloor)_{n \geq 1}$$

form a pair of complementary Beatty sequences. For other values of *p* the procedure for finding complementary sequences is not as straightforward.

Examples

For *r* = the golden mean, we have *s* = *r* + 1. In this case, the sequence $(\lfloor nr \rfloor)$, known as the *lower Wythoff sequence,* is

- 1, 3, 4, 6, 8, 9, 11, 12, 14, 16, 17, 19, 21, 22, 24, 25, 27, 29, ... (sequence A000201 in OEIS).

and the complementary sequence $(\lfloor ns \rfloor)$, the *upper Wythoff sequence,* is

- 2, 5, 7, 10, 13, 15, 18, 20, 23, 26, 28, 31, 34, 36, 39, 41, 44, 47, ... (sequence A001950 in OEIS).

These sequences define the optimal strategy for Wythoff's game. As another example, for *r* = "2, we have *s* = 2 + "2. In this case, the sequences are

- 1, 2, 4, 5, 7, 8, 9, 11, 12, 14, 15, 16, 18, 19, 21, 22, 24, ... (sequence A001951 in OEIS) and
- 3, 6, 10, 13, 17, 20, 23, 27, 30, 34, 37, 40, 44, 47, 51, 54, 58, ... (sequence A001952 in OEIS).

Notice that any number in the first sequence is lacking in the second, and vice versa.

History

Beatty sequences got their name from the problem posed in the *American Mathematical Monthly* by Samuel Beatty in 1926. It is probably one of the most often cited problems ever posed in the *Monthly*.

However, even earlier, in 1894 such sequences were briefly mentioned by John W. Strutt (3rd Baron Rayleigh) in the second edition of his book *The Theory of Sound*.

Rayleigh Theorem

The Rayleigh theorem (also known as Beatty's theorem) states that given an irrational $r > 1$, number there exists $s > 1$ so that the Beatty sequences $\mathcal{B}_r$ and $\mathcal{B}_s$ partition the set of positive integers: each positive integer belongs to exactly one of the two sequences.

First Proof

Given $r > 1$, let $s = r/(r-1)$. We must show that every positive integer lies in one and only one of the two sequences $\mathcal{B}_r$ and $\mathcal{B}_s$. We shall do so by considering the ordinal positions occupied by all the fractions j/r and k/s when they are jointly listed in nondecreasing order for positive integers j and k.

To see that no two of the numbers can occupy the same position (as a single number), suppose to the contrary that $j/r = k/s$ for some j and k. Then $r/s = j/k$, a rational number, but also, $r/s = r(1-1/r) = r-1$, not a rational number.

Therefore no two of the numbers occupy the same position.

For any j/r, there are j numbers $i/r \le j/r$ and $| js/r |$ numbers $, k/s \le j/r$ so that the position of j/r in the list is $j + | js/r |$. The equation $1/r + 1/s = 1$ implies

$$j + | js/r | = j + | j(s-1) | = | js |.$$

Likewise, the position of k/s in the list is $| kr |$.

Conclusion: every positive integer (that is, every position in the list) is of the form $| nr |$ or of the form , but not both.

The converse statement is also true: if p and q are two real numbers such that every positive integer occurs precisely once in the above list, then p and q are irrational and the sum of their reciprocals is 1.

Second Proof

Collisions: Suppose that, contrary to the theorem, there are integers $j > 0$ and k and m such that

$$j = \lfloor k{\cdot}r \rfloor = \lfloor m{\cdot}s \rfloor.$$

This is equivalent to the inequalities

$$j \le k{\cdot}r < j+1 \text{ and } j \le m{\cdot}s < j+1.$$

For non-zero j, the irrationality of r and s is incompatible with equality, so

$$j < k{\cdot}r < j+1 \text{ and } j < m{\cdot}s < j+1$$

which lead to

$$\frac{j}{r} < k < \frac{j+1}{r} \text{ and } \frac{j}{s} < m < \frac{j+1}{s}.$$

Adding these together and using the hypothesis, we get

$$j < k+m < j+1$$

which is impossible (one cannot have an integer between two adjacent integers). Thus the supposition must be false.

Anti-collisions: Suppose that, contrary to the theorem, there are integers $j > 0$ and k and m such that

$$k{\cdot}r < j \text{ and } j+1 \le (k+1){\cdot}r \text{ and } m{\cdot}s < j \text{ and } j+1 \le (m+1){\cdot}s.$$

Since $j + 1$ is non-zero and r and s are irrational, we can exclude equality, so

$$k{\cdot}r < j \text{ and } j+1 < (k+1){\cdot}r \text{ and } m{\cdot}s < j \text{ and } j+1 < (m+1){\cdot}s.$$

Then we get

$$k < \frac{j}{r} \text{ and } \frac{j+1}{r} < k+1 \text{ and } m < \frac{j}{s} \text{ and } \frac{j+1}{s} < m+1$$

Adding corresponding inequalities, we get

$$k+m < j \text{ and } j+1 < k+m+2$$

$$k+m < j < k+m+1$$

which is also impossible. Thus the supposition is false.

Properties

1. $m \in \mathcal{B}_r$ if and only

$$0 \leq 1 - \frac{1}{r} \leq \left[\frac{m}{r} \right]_1$$

where $[x]_1$ denotes $x \mod 1$ or the fractional part of xi.e., $[x]_1 = x - \lfloor x \rfloor$. Furthermore, if $m \in \mathcal{B}_r$

$$m = \left\lfloor \left(\left\lfloor \frac{m}{r} \right\rfloor + 1 \right) r \right\rfloor$$

Proof: $m = \left\lfloor \frac{m}{r} \right\rfloor r + \left[\frac{m}{r} \right]_1 r = \left(\left\lfloor \frac{m}{r} \right\rfloor + 1 \right) r - \left(1 - \left[\frac{m}{r} \right]_1 \right) r$

If $\left\lfloor \left(\left\lfloor \frac{m}{r} \right\rfloor + 1 \right) r \right\rfloor = m$, then $\left\lfloor \left(\frac{m}{r} - \left[\frac{m}{r} \right]_1 + 1 \right) r \right\rfloor = m$

Or, $\left\lfloor m + \left(1 - \left[\frac{m}{r} \right]_1 \right) r \right\rfloor = m$ and thus, $0 \leq \left(1 - \left[\frac{m}{r} \right]_1 \right) r < 1$ which is the given condition rearranged.

Relation with Sturmian Sequences

The first difference

$$\lfloor (n+1)r \rfloor - \lfloor nr \rfloor$$

of the Beatty sequence associated to the irrational number r is a characteristic Sturmian word over the alphabet $\{\lfloor r \rfloor, \lfloor r \rfloor + 1\}$.

Generalizations

The Lambek–Moser theorem generalizes the Rayleigh theorem and shows that more general pairs of sequences defined from an integer function and its inverse have the same property of partitioning the integers.

Birch's Theorem

In mathematics, Birch's theorem, named for Bryan John Birch, is a statement about the representability of zero by odd degree forms.

Statement of Birch's Theorem

Let K be an algebraic number field, k, l and n be natural numbers, $r_1,...,r_k$ be odd natural numbers, and $f_1,...,f_k$ be homogeneous polynomials with coefficients in K of degrees $r_1,...,r_k$ respectively in n variables, then there exists a number $\psi(r_1,...,r_k,l,K)$ such that

$$n \geq \psi(r_1,\ldots,r_k,l,K)$$

implies that there exists an l-dimensional vector subspace V of K^n such that

$$f_1(x) = \ldots f_k(x) = 0, \quad \forall x \in V.$$

Remarks

The proof of the theorem is by induction over the maximal degree of the forms $f_1,...,f_k$. Essential to the proof is a special case, which can be proved by an application of the Hardy-Littlewood circle method, of the theorem which states that if n is sufficiently large and r is odd, then the equation

$$c_1x_1^r + \ldots + c_nx_n^r = 0, \quad c_i \in \mathbb{Z}, i = 1,\ldots,n$$

has a solution in integers $x_1,...,x_n$, not all of which are 0.

The restriction to odd r is necessary, since even-degree forms, such as positive definite quadratic forms, may take the value 0 only at the origin.

Brauer's Theorem on Forms

In mathematics, Brauer's theorem, named for Richard Brauer, is a result on the representability of 0 by forms over certain fields in sufficiently many variables.

Statement of Brauer's Theorem

Let K be a field such that for every integer $r > 0$ there exists an integer $\psi(r)$ such that for n e– ψ(r) every equation

$$(*) \qquad a_1x_1^r + \cdots + a_nx_n^r = 0, \quad a_i \in K, \quad i = 1,\ldots,n$$

has a non-trivial (i.e. not all x_i are equal to 0) solution in K. Then, given homogeneous polynomials $f_1,...,f_k$ of degrees $r_1,...,r_k$ respectively with coefficients in K, for every set of positive integers

$r_1,...,r_k$ and every non-negative integer l, there exists a number $\omega(r_1,...,r_k,l)$ such that for n e– $\omega(r_1,...,r_k,l)$ there exists an l-dimensional affine subspace M of K^n (regarded as a vector space over K) satisfying

$$f_1(x_1,\ldots,x_n)=\cdots=f_k(x_1,\ldots,x_n)=0,\quad \forall(x_1,\ldots,x_n)\in M.$$

An application to the Field of p-adic Numbers

Letting K be the field of p-adic numbers in the theorem, the equation (*) is satisfied, since $\mathbb{Q}_p^* / \left(\mathbb{Q}_p^*\right)^b$, b a natural number, is finite. Choosing $k = 1$, one obtains the following corollary:

A homogeneous equation $f(x_1,...,x_n) = 0$ of degree r in the field of p-adic numbers has a non-trivial solution if n is sufficiently large.

One can show that if n is sufficiently large according to the above corollary, then n is greater than r^2. Indeed, Emil Artin conjectured that every homogeneous polynomial of degree r over Q_p in more than r^2 variables represents 0. This is obviously true for $r = 1$, and it is well-known that the conjecture is true for $r = 2$.

In 1950 Demyanov verified the conjecture for $r = 3$ and $p \neq 3$, and in 1952 D. J. Lewis independently proved the case $r = 3$ for all primes p. But in 1966 Guy Terjanian constructed a homogeneous polynomial of degree 4 over Q_2 in 18 variables that has no non-trivial zero. On the other hand, the Ax–Kochen theorem shows that for any fixed degree Artin's conjecture is true for all but finitely many Q_p.

Carmichael's Theorem

Carmichael's theorem, named after the American mathematician R.D. Carmichael, states that for n greater than 12, the nth Fibonacci number F(n) has at least one prime divisor that does not divide any earlier Fibonacci number.

The only exceptions for n up to 12 are:

F(1)=1 and F(2)=1, which have no prime divisors

F(6)=8 whose only prime divisor is 2 (which is F(3))

F(12)=144 whose only prime divisors are 2 (which is F(3)) and 3 (which is F(4))

If a prime p is a divisor of $F(n)$ that does not divide any $F(m)$ with $m < n$, then p is called a *characteristic factor* or a *primitive prime divisor* of $F(n)$. Carmichael's theorem says that every Fibonacci number, apart from the exceptions listed above, has at least one primitive prime divisor.

The theorem can be generalized from Fibonacci numbers to other Lucas sequences.

Chebyshev's Bias

In number theory, Chebyshev's bias is the phenomenon that most of the time, there are more primes of the form $4k + 3$ than of the form $4k + 1$, up to the same limit. This phenomenon was first observed by Chebyshev in 1853.

Description

Let $\pi(x; 4, 1)$ denote the number of primes of the form $4k + 1$ up to x. Similarly, let $\pi(x; 4, 3)$ denote the number of primes of the form $4k + 3$ up to x. By the prime number theorem, extended to arithmetic progression,

$$\pi(x;4,1) \sim \pi(x;4,3) \sim \frac{1}{2}\frac{x}{\log x},$$

i.e., half of the primes are of the form $4k+1$, and half of the form $4k + 3$. A reasonable guess would be that $\pi(x;4,1)>\pi(x; 4, 3)$ and $\pi(x; 4, 1) < \pi(x; 4, 3)$ each also occur 50 % of the time. This, however, is not supported by numerical evidence—in fact, $\pi(x,4,3)>\pi(x;4,1)$ occurs much more frequently.

In general, if $0 < a,b < q$ are integers, $(a, q) = (b, q) = 1$, a is a quadratic residue, b is a quadratic nonresidue mod q, then $\pi(x; q, b) > \pi(x; q, a)$ occurs more often than not. This has been proved only by assuming strong forms of the Riemann hypothesis. The conjecture of Knapowski and Turan, however, that the density of the numbers x for which $\pi(x; 4, 3) > \pi(x; 4, 1)$ holds, is 1, turned out to be false. They, however, do have a logarithmic density, which is approximately 0.9959...(Rubinstein—Sarnak, 1994).

Chinese Remainder Theorem

The Chinese remainder theorem is a result about congruences in number theory and its generalizations in abstract algebra. In its basic form, the Chinese remainder theorem will determine a number n that when divided by some given divisors leaves given remainders.

For example, what is the lowest number n that when divided by 3 leaves a remainder of 2, when divided by 5 leaves a remainder of 3, and when divided by 7 leaves a remainder of 2? A common introductory example is a woman who tells a policeman that she lost her basket of eggs, and that if she took three at a time out of it, she was left with 2, if she took five at a time out of it she was left with 3, and if she took seven at a time out of it she was left with 2. She then asks the policeman what is the minimum number of eggs she must have had. The answer to both problems is 23.

Theorem Statement

The original form of the theorem, contained in a third-century AD book *Sun Zi suanjing* by Chinese mathematician Sun Tzu and later republished in a 1247 book by Qin Jiushao, the *Shushu Jiuzhang* is a statement about simultaneous congruences.

Suppose $n_1, n_2, \ldots, n_k$ are positive integers which are pairwise coprime. Then, for any given sequence of integers $a_1, a_2, \ldots, a_k$, there exists an integer x solving the following system of simultaneous congruences.

$$\begin{aligned} x &\equiv a_1 \pmod{n_1} \\ x &\equiv a_2 \pmod{n_2} \\ &\vdots \\ x &\equiv a_k \pmod{n_k} \end{aligned}$$

Furthermore, all solutions x of this system are congruent modulo the product, $N = n_1 n_2 \ldots n_k$.

Hence $x \equiv y \pmod{n_i}$ for all $1 \le i \le k$, if and only if $x \equiv y \pmod{N}$.

Sometimes, the simultaneous congruences can be solved even if the n_i's are not pairwise coprime. A solution x exists if and only if:

$$a_i \equiv a_j \pmod{\gcd(n_i, n_j)} \qquad \text{for all } i \text{ and } j.$$

All solutions x are then congruent modulo the least common multiple of the n_i.

Sun Zi's work contains neither a proof nor a full algorithm. What amounts to an algorithm for solving this problem was described by Aryabhata (6th century). Special cases of the Chinese remainder theorem were also known to Brahmagupta (7th century), and appear in Fibonacci's Liber Abaci (1202).

A modern restatement of the theorem in algebraic language is that for a positive integer $n$with prime factorization $p_1^{r_1} p_2^{r_2} \cdots p_k^{r_k}$ we have the isomorphism between a ring and the direct product of its prime power parts:

$$\mathbb{Z} / n\mathbb{Z} \cong \mathbb{Z} / p_1^{r_1}\mathbb{Z} \times \mathbb{Z} / p_2^{r_2}\mathbb{Z} \times \cdots \times \mathbb{Z} / p_k^{r_k}\mathbb{Z}.$$

Existence

Existence can be seen by an explicit construction of x. We will use the notation $[a^{-1}]_b$ to denote the multiplicative inverse of $a \pmod{b}$, it is defined exactly when aand bare coprime; the following construction explains why the coprimality condition is needed.

Case of Two Equations

Given the system (corresponding to $k=2$)

$$\begin{aligned} x &\equiv a_1 \pmod{n_1} \\ x &\equiv a_2 \pmod{n_2}. \end{aligned}$$

Since $\gcd(n_1, n_2)=1$, we have from Bézout's identity

$$n_2[n_2^{-1}]_{n_1} + n_1[n_1^{-1}]_{n_2} = 1.$$

Multiplying both sides by x, we get

$$x = xn_2[n_2^{-1}]_{n_1} + xn_1[n_1^{-1}]_{n_2}.$$

If we take the congruence modulo n_1 for the right-hand-side expression, it is readily seen that

$$xn_2[n_2^{-1}]_{n_1} + xn_1[n_1^{-1}]_{n_2} \equiv x \times 1 + x \times 0 \times [n_1^{-1}]_{n_2} \equiv x \pmod{n_1}.$$

But we know that

$$x \equiv a_1 \pmod{n_1},$$

thus this suggests that the coefficient of the first term on the right-hand-side expression should be a_1. Similarly, we can show that the coefficient of the second term should be a_2.

We can now define the value

$$x := a_1 n_2 [n_2^{-1}]_{n_1} + a_2 n_1 [n_1^{-1}]_{n_2}$$

and it is seen to satisfy both congruences by reducing. For example

$$a_1 n_2 [n_2^{-1}]_{n_1} + a_2 n_1 [n_1^{-1}]_{n_2} \equiv a_1 1 + a_2 0 [n_1^{-1}]_{n_2} \equiv a_1 \pmod{n_1}.$$

General Ease

The same type of construction works in the general case of congruence equations. Let $N = n_1 n_2 \cdots n_k$ be the product of every modulus then define

$$x := \sum_i a_i \frac{N}{n_i} \left[\left(\frac{N}{n_i} \right)^{-1} \right]_{n_i}$$

and this is seen to satisfy the system of congruences by a similar calculation as before.

Finding the Solution with Basic Algebra and Modular Arithmetic

For example, consider the problem of finding an integer x such that

$$x \equiv 2 \pmod{3}$$
$$x \equiv 3 \pmod{4}$$
$$x \equiv 1 \pmod{5}.$$

A brute-force approach converts these congruences into sets and writes the elements out to the product of 3×4×5 = 60 (the solutions modulo 60 for each congruence):

x∈ {2, 5, 8, 11, 14, 17, 20, 23, 26, 29, 32, 35, 38, 41, 44, 47, 50,

53, 56, 59, ...}

x∈ {3, 7, 11, 15, 19, 23, 27, 31, 35, 39, 43, 47, 51, 55, 59, ...}

x∈ {1, 6, 11, 16, 21, 26, 31, 36, 41, 46, 51, 56, ...}.

To find an x that satisfies all three congruences, intersect the three sets to get:

x∈ {11, ...}.

Which can be expressed as

$$x \equiv 11 \pmod{60}.$$

Another way to find a solution is with basic algebra, modular arithmetic, and stepwise substitution.

We start by translating these equivalences into equations for some t, s, and u:

- Equation 1: x = 2 + 3 × t (mod 3)
- Equation 2: x = 3 + 4 × s (mod 4)
- Equation 3: x = 1 + 5 × u (mod 5).

Start by substituting the x from equation 1 into equivalence 2: 2 + 3 × t = 3 (mod 4), hence 3 × t = 1 (mod 4), or t = (1/3) (mod 4) = 3 (mod 4), meaning that t = 3 + 4 × s for integer s.

Plug t into equation 1: x = 2 + 3 × t (mod 3) = 2 + 3 × (3 + 4 × s) (mod 3) = 11 + 12 × s (mod 3).

Plug this x into equivalence 3: 11 + 12 × s = 1 (mod 5). Casting out 5s, we get 1 + 2 × s = 1 (mod 5), or 2 × s = 0 (mod 5), meaning that s = 0 + 5 × u for integer u.

Finally, x = 11 + 12 × s = 11 + 12 × (5 × u) = 11 + (60 × u). Since 60 = lcm(3, 4, 5), we have solutions 11, 71, 131, 191, ...

A Constructive Algorithm to Find the Solution

The following algorithm only applies if the n_i's are pairwise coprime. (For simultaneous congruences when the moduli are not pairwise coprime, the method of successive substitution can often yield solutions.)

Suppose, as above, that a solution is required for the system of congruences:

$$x \equiv a_i \pmod{n_i} \quad \text{for } i = 1, \ldots, k.$$

Again, to begin, the product $N = n_1 n_2 \ldots n_k$ is defined. Then a solution x can be found as follows.

For each i the integers n_i and N / n_i are coprime. Using the extended Euclidean algorithm we can find integers r_i and s_i such that $r_i n_i + s_i N / n_i = 1$. Then, choosing the label , the above expression becomes:

$$r_i n_i + e_i = 1.$$

Consider e_i. The above equation guarantees that its remainder, when divided by n_i, must be 1. On the other hand, since it is formed as $s_i N / n_i$, the presence of N guarantees that it's evenly divisible by any n_j so long as $j \neq i$.

$$e_i \equiv 1 \pmod{n_i} \quad \text{and} \quad e_i \equiv 0 \pmod{n_j} \quad \text{for } j \neq i$$

Because of this, combined with the multiplication rules allowed in congruences, one solution to the system of simultaneous congruences is:

$$x = \sum_{i=1}^{k} a_i e_i.$$

For example, consider the problem of finding an integer x such that

$$\begin{aligned} x &\equiv 2 \pmod{3} \\ x &\equiv 3 \pmod{4} \\ x &\equiv 1 \pmod{5}. \end{aligned}$$

Using the extended Euclidean algorithm, for x modulo 3 and 20 [4×5], we find (-13) × 3 + 2 × 20 = 1, i.e. e_1 = 40. For x modulo 4 and 15 [3×5], we get (-11) × 4 + 3 × 15 = 1, i.e. e_2 = 45. Finally, for x modulo 5 and 12 [3×4], we get 5 × 5 + (-2) × 12 = 1, i.e. e_3 = -24. A solution x is therefore 2 × 40 + 3 × 45 + 1 × (-24) = 191. All other solutions are congruent to 191 modulo 60, [3 × 4 × 5 = 60], which means they are all congruent to 11 modulo 60.

Note: There are multiple implementations of the extended Euclidean algorithm which will yield different sets of $e_1 = -20$, $e_2 = -15$, and $e_3 = -24$. These sets however will produce the same solution; i.e., (-20)2 + (-15)3 + (-24)1 = -109 = 11 modulo 60.

Statement for Principal Ideal Domains

For a principal ideal domain *R* the Chinese remainder theorem takes the following form: If $u_1, \ldots, u_k$ are elements of *R* which are pairwise coprime, and *u* denotes the product $u_1 \ldots u_k$, then the quotient ring *R*/*uR* and the product ring $R/u_1R \times \ldots \times R/u_kR$ are isomorphic via the isomorphism

$$f : R/uR \to R/u_1R \times \cdots \times R/u_kR$$

such that

$$f(x+uR) = (x+u_1R, \ldots, x+u_kR) \quad \text{for every } x \in R.$$

This map is well-defined and an isomorphism of rings; the inverse isomorphism can be constructed as follows. For each *i*, the elements u_i and u/u_i are coprime, and therefore there exist elements *r* and *s* in *R* with

$$ru_i + su/u_i = 1.$$

Set $e_i = s\, u/u_i$. Then the inverse of *f* is the map

$$g : R/u_1R \times \cdots \times R/u_kR \to R/uR$$

such that

$$g(a_1 + u_1R, \ldots, a_k + u_kR) = \left(\sum_{i=1}^{k} a_i \frac{u}{u_i} \left[\left(\frac{u}{u_i} \right)^{-1} \right]_{u_i} \right) + uR \quad \text{for all } a_1, \ldots, a_k \in R.$$

Note that this statement is a straightforward generalization of the above theorem about integer congruences: the ring Z of integers is a principal ideal domain, the surjectivity of the map *f* shows that every system of congruences of the form

$$x \equiv a_i \pmod{u_i} \quad \text{for } i = 1, \ldots, k$$

can be solved for *x*, and the injectivity of the map *f* shows that all the solutions *x* are congruent modulo *u*.

Statement for General Rings

The general form of the Chinese remainder theorem, which implies all the statements given above, can be formulated for commutative rings and ideals. If *R* is a commutative ring and I_1, ..., I_k are ideals of *R* which are pairwise coprime (meaning that $I_i + I_j = R$ for all) $i \neq j$, then the product *I* of these ideals is equal to their intersection, and the quotient ring *R/I* is isomorphic to the product ring $R/I_1 \times R/I_2 \times \ldots \times R/I_k$ via the isomorphism

$$f : R/I \to R/I_1 \times \cdots \times R/I_k$$

such that

$$f(x+I) = (x+I_1, \ldots, x+I_k) \quad \text{for all } x \in R.$$

Here is a version of the theorem where *R* is not required to be commutative:

Let *R* be any ring with 1 (not necessarily commutative) and $I_1, \ldots, I_n$ be pairwise coprime 2-sided ideals. Then the canonical R-module homomorphism $R \to R/I_1 \times \cdots \times R/I_k$ is onto, with kernel $I_1 \cap \cdots \cap I_k$. Hence, $R/(I_1 \cap \cdots \cap I_k) \simeq R/I_1 \times \cdots \times R/I_k$ (as *R*-modules).

Applications

- In the RSA algorithm calculations are made modulo *n*, where *n* is a product of two large prime numbers *p* and *q*. 1,024-, 2,048- or 4,096-bit integers *n* are commonly used, making calculations $\mathbb{Z}/n\mathbb{Z}$ in very time-consuming. By the Chinese remainder theorem, however, these calculations can be done in the isomorphic ring $\mathbb{Z}/p\mathbb{Z} \oplus \mathbb{Z}/q\mathbb{Z}$ instead. Since *p* and *q* are normally of about $\sqrt{n}$ the same size, that is about , calculations in the latter representation are much faster. Note that RSA algorithm implementations using this isomorphism are more susceptible to fault injection attacks.
- The Chinese remainder theorem may also be used to construct an elegant Gödel numbering for sequences, which is needed to prove Gödel's incompleteness theorems.

- The following example shows a connection with the classic polynomial interpolation theory. Let r complex points ("interpolation nodes") $\lambda_1,\ldots,\lambda_r$ be given, together with the complex data $a_{j,k}$, for all $1\le j\le r$ and $0\le k<\nu_j$. The general Hermite interpolation problem asks for a polynomial $P(x)\in \mathbb{C}[x]$ taking the prescribed derivatives in each node λ_j:

$$P^{(k)}(\lambda_j)=a_{j,k} \quad \forall 1\le j\le r, 0\le k<\nu_j.$$

Introducing the polynomials

$$A_j(x):=\sum_{k=0}^{\nu_j-1}\frac{a_{j,k}}{k!}(x-\lambda_j)^k$$

the problem may be equivalently reformulated as a system of simultaneous congruences:

$$P(x)\equiv A_j(x) \quad \left(\operatorname{mod}(x-\lambda_j)^{\nu_j}\right), \quad \forall 1\le j\le r.$$

By the Chinese remainder theorem in the principal ideal domain $\mathbb{C}[x]$, there is a unique such polynomial $P(x)$ with degree $\deg(P)<n:=\sum_j \nu_j$. A direct construction, in analogy with the above proof for the integer number case, can be performed as follows. Define the polynomials $Q:=\prod_{i=1}^{r}(x-\lambda_i)^{\nu_i}$ and $Q_j:=\dfrac{Q}{(x-\lambda_j)^{\nu_j}}$. The partial fraction decomposition of $\dfrac{1}{Q}$ gives r polynomials S_j with degrees $\deg(S_j)<\nu_j$ such that

$$\frac{1}{Q}=\sum_{i=1}^{r}\frac{S_i}{(x-\lambda_i)^{\nu_i}}$$

so that $1=\sum_{i=1}^{r}S_iQ_i$. Then a solution of the simultaneous congruence system is given by the polynomial

$$\sum_{i=1}^{r} A_i S_i Q_i = A_j + \sum_{i=1}^{r} (A_i - A_j) S_i Q_i \equiv A_j \quad \left(\bmod (x-\lambda_j)^{\nu_j}\right) \quad \forall 1 \le j \le r$$

and the minimal degree solution is this one reduced modulo Q, that is the unique with degree less than n.

- The Chinese remainder theorem can also be used in secret sharing, which consists of distributing a set of shares among a group of people who, all together (but no one alone), can recover a certain secret from the given set of shares. Each of the shares is represented in a congruence, and the solution of the system of congruences using the Chinese remainder theorem is the secret to be recovered. Secret Sharing using the Chinese Remainder Theorem uses, along with the Chinese remainder theorem, special sequences of integers that guarantee the impossibility of recovering the secret from a set of shares with less than a certain cardinality.
- The Good-Thomas fast Fourier transform algorithm exploits a re-indexing of the data based on the Chinese remainder theorem. The Prime-factor FFT algorithm contains an implementation.
- Dedekind's theorem on the linear independence of characters states (in one of its most general forms) that if M is a monoid and k is an integral domain, then any finite family $(f_i)_{i \in I}$ of distinct monoid homomorphisms $f_i : M \to k$ (where the monoid structure on k is given by multiplication) is linearly independent; i.e., every family $(\alpha_i)_{i \in I}$ of elements $\alpha_i \in k$ satisfying $\sum_{i \in I} \alpha_i f_i = 0$ must be equal to the family $(0)_{i \in I}$.

Proof using the Chinese Remainder Theorem: First, assume that k is a field (otherwise, replace the integral domain k by its quotient field, and nothing will change). We can linearly extend the monoid homomorphisms $f_i : M \to k$ to k-algebra homomorphisms $F_i : k[M] \to k$, where $k[M]$ is the monoid ring of M over k.

Then, the condition $\sum_{i \in I} \alpha_i f_i = 0$ yields $\sum_{i \in I} \alpha_i F_i = 0$ by linearity. Now, we notice that if $i \neq j$ are two elements of the index set I, then the two k-linear maps $F_i : k[M] \to k$ and $F_j : k[M] \to k$ are not proportional to each other (because if they were, then f_i and f_j would also be proportional to each other, and thus equal to each other since $f_i(1) = 1 = f_j(1)$ (since f_i and f_j are monoid homomorphisms), contradicting the assumption that they be distinct). Hence, their kernels $\operatorname{Ker} F_i$ and are $\operatorname{Ker} F_j$ distinct. Now, $\operatorname{Ker} F_i$ is a maximal ideal of $k[M]$ for every $i \in I$ (since $k[M] / \operatorname{Ker} F_i \cong F_i(k[M]) = k$ is a field), and the ideals $\operatorname{Ker} F_i$ and $\operatorname{Ker} F_j$ are coprime whenever $i \neq j$ (since they are distinct and maximal). The Chinese Remainder Theorem (for general rings)) thus yields that the map

$$\phi : k[M] / K \to \prod_{i \in I} k[M] / \operatorname{Ker} F_i$$

given by

$$\phi(x + K) = (x + \operatorname{Ker} F_i)_{i \in I} \text{ for all } x \in k[M]$$

is an isomorphism, where $K = \prod_{i \in I} \operatorname{Ker} F_i = \bigcap_{i \in I} \operatorname{Ker} F_i$.

Consequently, the map

$$\Phi : k[M] \to \prod_{i \in I} k[M] / \operatorname{Ker} F_i$$

given by

$$\Phi(x) = (x + \operatorname{Ker} F_i)_{i \in I} \text{ for all } x \in k[M]$$

is surjective. Under the isomorphisms $k[M] / \operatorname{Ker} F_i \to F_i(k[M]) = k$, this map Φ corresponds to the map

$$\psi : k[M] \to \prod_{i \in I} k$$

given by

$$x \mapsto \left[F_i(x)\right]_{i\in I} \text{ for every } x \in k[M].$$

Now, $\sum_{i\in I} \alpha_i F_i = 0$ yields $\sum_{i\in I} \alpha_i u_i = 0$ for every vector $(u_i)_{i\in I}$ in the image of the map ψ. Since ψ is surjective, this means $\sum_{i\in I} \alpha_i u_i = 0$ that for every vector $(u_i)_{i\in I} \in \prod_{i\in I} k$. Consequently, $(\alpha_i)_{i\in I} = (0)_{i\in I}$, QED.

Non-commutative Case: a Counter-example

The Chinese remainder theorem does not hold in the non-commutative case. Consider the ring *R* of non-commutative real polynomials in *x* and *y*. Let *I* be the principal two-sided ideal generated by *x* and *J* the principal two-sided ideal generated by $xy+1$. Then $I+J=R$ but $I \cap J \neq IJ$.

Proof

Observe that *I* is formed by all polynomials with an *x* in every term and that every polynomial in *J* vanishes under the substitution $y=-1/x$. Consider the polynomial $p=(xy+1)x$. Clearly $p \in I \cap J$. Define a term in *R* as an element of the multiplicative monoid of *R* generated by *x* and *y*. Define the degree of a term as the usual degree of the term after the substitution $y=x$. On the other hand, suppose $q \in J$. Observe that a term in *q* of maximum degree depends on *y* otherwise *q* under the substitution $y=-1/x$ can not vanish. The same happens then for an element $q \in IJ$. Observe that the last *y*, from left to right, in a term of maximum degree in an element of *IJ* is preceded by more than one *x*. (We are counting here all the preceding *x*s. E.g., in x^2yxyx^5 the last *y* is preceded by 3*x*s.) This proves that $(xy+1)x \notin IJ$ since that last *y* in a term of maximum degree (xyx) is preceded by only one *x*. Hence $I \cap J \neq IJ$.

On the other hand, it is true in general that $I+J=R$ implies $I \cap J = IJ + JI$.

To see this, note that $I \cap J = (I \cap J)(I+J) \subset IJ + JI$, while the opposite inclusion is obvious. Also, we have in general that, provided $I_1, \ldots, I_m$ are pairwise coprime two-sided ideals in *R*, the natural map

$$R/(I_1 \cap I_2 \cap \ldots \cap I_m) \rightarrow R/I_1 \oplus R/I_2 \oplus \cdots \oplus R/I_m$$

is an isomorphism. Note that $I_1 \cap I_2 \cap \ldots \cap I_m$ can be replaced by a sum over all orderings of $I_1, \ldots, I_m$ of their product (or just a sum over enough orderings, using inductively that $I \cap J = IJ + JI$ for coprime ideals I, J).

Chowla–Mordell Theorem

In mathematics, the Chowla–Mordell theorem is a result in number theory determining cases where a Gauss sum is the square root of a prime number, multiplied by a root of unity. It was proved and published independently by Sarvadaman Chowla and Louis Mordell, around 1951.

In detail, *P* if is a prime number, X a nontrivial Dirichlet character modulo *P*, and

$$G(\chi) = \sum \chi(a) \zeta^a$$

where ξis a primitive *P*-th root of unity in the complex numbers, then

$$\frac{G(\chi)}{|G(\chi)|}$$

is a root of unity if and only if Xis the quadratic residue symbol modulo *P*. The 'if' part was known to Gauss: the contribution of Chowla and Mordell was the 'only if' direction. The ratio in the theorem occurs in the functional equation of L-functions.

Cubic Reciprocity

Cubic reciprocity is a collection of theorems in elementary and algebraic number theory that state conditions under which the congruence x^3 a= p (mod q) is solvable; the word "reciprocity" comes from the form of the main theorem, which states that if p and q are primary numbers in the ring of Eisenstein integers, both coprime to 3,

The congruence x^3 a– p (mod q) is solvable if and only if x^3 a– q (mod p) is.

History

Sometime before 1748 Euler made the first conjectures about the cubic residuacity of small integers, but they were not published until 1849, after his death.

Gauss's published works mention cubic residues and reciprocity three times: there is one result pertaining to cubic residues in the Disquisitiones Arithmeticae (1801). In the introduction to the fifth and sixth proofs of quadratic reciprocity (1818) he said that he was publishing these proofs because their techniques (Gauss's lemma and Gaussian sums, respectively) can be applied to cubic and biquadratic reciprocity. Finally, a footnote in the second (of two) monographs on biquadratic reciprocity (1832) states that cubic reciprocity is most easily described in the ring of Eisenstein integers.

From his diary and other unpublished sources, it appears that Gauss knew the rules for the cubic and quartic residuacity of integers by 1805, and discovered the full-blown theorems and proofs of cubic and biquadratic reciprocity around 1814. Proofs of these were found in his posthumous papers, but it is not clear if they are his or Eisenstein's.

Jacobi published several theorems about cubic residuacity in 1827, but no proofs. In his Königsberg lectures of 1836–37 Jacobi presented proofs. The first published proofs were by Eisenstein (1844).

Integers

A cubic residue (mod p) is any number congruent to the third power of an integer (mod p). If x^3 a– a (mod p) does not have an integer solution, a is a cubic nonresidue (mod p).

As is often the case in number theory, it is easiest to work modulo prime numbers, so in this section all moduli p, q, etc., are assumed to be positive, odd primes.

The first thing to notice when working within the ring Z of integers is that if the prime number q is a– 2 (mod 3) every number is a cubic residue (mod q). Let $q = 3n + 2$; since $0 = 0^3$ is obviously a cubic residue, assume x is not divisible by q. Then by Fermat's little theorem,

$$x^q = x^{3n+2} \equiv x \pmod{q} \text{ and } x^{q-1} = x^{3n+1} \equiv 1 \pmod{q}, \text{ so}$$

$$x = 1{\cdot}x \equiv x^q x^{q-1} = x^{3n+2} x^{3n+1} = x^{6n+3} = (x^{2n+1})^3 \pmod{q}$$

is a cubic residue (mod q).

Therefore, the only interesting case is when the modulus p a– 1 (mod 3).

In this case, p a– 1 (mod 3), the nonzero residue classes (mod p) can be divided into three sets, each containing $(p-1)/3$ numbers. Let e be a cubic nonresidue.

The first set is the cubic residues; the second one is e times the numbers in the first set, and the third is e^2 times the numbers in the first set.

Another way to describe this division is to let e be a primitive root (mod p); then the first (respectively second, third) set is the numbers whose indices with respect to this root are a– 0 (resp. 1, 2) (mod 3).

In the vocabulary of group theory, the first set is a subgroup of index 3 (of the multiplicative group $Z/pZ^{\times}$), and the other two are its cosets.

Primes a– 1 (mod 3)

A theorem of Fermat states that every prime p a– 1 (mod 3) is the sum of a square and three times a square: $p = a^2 + 3b^2$ and (except for the signs of a and b) this representation is unique.

Letting $m = a + b$ and $n = a - b$, we see that this is equivalent to $p = m^2 - mn + n^2$ (which equals $(n - m)^2 - (n - m)n + n^2 = m^2 + m(n - m) + (n - m)^2$, so m and n are not determined uniquely). Thus,

$$\begin{aligned} 4p &= (2m-n)^2 + 3n^2 \\ &= (2n-m)^2 + 3m^2 \\ &= (m+n)^2 + 3(m-n)^2, \end{aligned}$$

and it is a straightforward exercise to show that exactly one of m, n, or $m - n$ is a multiple of 3, so

$p = \frac{1}{4}\left(L^2 + 27M^2\right)$, and this representation is unique up to the signs of L and M.

For relatively-prime integers m and n define the rational cubic residue symbol as

$$\left[\frac{m}{n}\right]_3 = \begin{cases} +1 \text{ if } m \text{ is a cubic residue} \pmod{n} \\ -1 \text{ if } m \text{ is a cubic nonresidue} \pmod{n} \end{cases}$$

Euler

Euler's conjectures are based on the representation $p = 3a^2 + b^2$. The symbol $m \mid n$ is read "m divides n" and means there is an a such that $n = ma$.

$\left[\frac{2}{p}\right]_3 = 1$ if and only if $3 \mid a$

$\left[\frac{3}{p}\right]_3 = 1$ if and only if $9 \mid a$; or $9 \mid (a \pm b)$

$\left[\frac{5}{p}\right]_3 = 1$ if and only if $15 \mid a$; or $3 \mid a$ and $5 \mid b$; or $15 \mid (a \pm b)$; or $15 \mid (a \pm 2b)$

$\left[\frac{6}{p}\right]_3 = 1$ if and only if $9 \mid a$; or $9 \mid (2a \pm b)$

$\left[\frac{7}{p}\right]_3 = 1$ if and only if $21 \mid a$; or $3 \mid a$ and $7 \mid b$; or $21 \mid (a \pm b)$; or $7 \mid (4a \pm b)$; or $7 \mid (a \pm 2b)$

The first two can be restated as

Let p a– 1 (mod 3) be a positive prime. Then 2 is a cubic residue of p if and only if $p = a^2 + 27b^2$.

Let p a– 1 (mod 3) be a positive prime. Then 3 is a cubic residue of p if and only if $4p = a^2 + 243b^2$.

Gauss

Gauss provs that if $p = 3n + 1 = \frac{1}{4}\left(L^2 + 27M^2\right)$, then $L(n!)^3 \equiv 1 \pmod{p}$, from which $\left[\frac{L}{p}\right]_3 = \left[\frac{M}{p}\right]_3 = 1$ is an easy deduction.

Jacobi

Jacobi stated (without proof)

Let q a= p a= 1 (mod 6) be positive primes, $p = \frac{1}{4}\left(L^2 + 27M^2\right)$, and let x be a solution of x^2 a– "3 (mod q). Then

$$\left[\frac{q}{p}\right]_3 = 1 \text{ if and only if } \left[\frac{\frac{L+3Mx}{2}p}{q}\right]_3 = 1 \text{ if and only if } \left[\frac{(\frac{L+3Mx}{L-3Mx})}{q}\right]_3 = 1.$$

(The "numerator" in the last expression is an integer (mod q), not a Legendre symbol).

If $q = \frac{1}{4}\left(L'^2 + 27M'^2\right)$, then $x \equiv \pm\frac{L'}{3M'} \pmod{q}$, and we have

$$\left[\frac{q}{p}\right]_3 = 1 \text{ if and only if } \left[\frac{(\frac{LM' + L'M}{LM' - L'M})}{p}\right]_3 = 1.$$

Along the same lines, von Lienen proved

$$\left[\frac{p}{q}\right]_3 \left[\frac{q}{p}\right]_3 = \left[\frac{(\frac{LM' + L'M}{2M})}{q}\right]_3^2.$$

Other Theorems

Emma Lehmer proved

Let q and $p = \frac{1}{4}\left(L^2 + 27M^2\right)$ be primes.

$$\left[\frac{q}{p}\right]_3 = 1 \text{ if and only if } \begin{cases} q \mid LM \text{ or} \\ L \equiv \pm\dfrac{9r}{2u+1}M \pmod{q}, \quad \text{where} \\ u \not\equiv 0, 1, -\dfrac{1}{2}, -\dfrac{1}{3} \pmod{q} \quad \text{and} \\ 3u + 1 \equiv r^2(3u - 3) \pmod{q} \end{cases}$$

Note that the first condition implies:

Any number that divides L or M is a cubic residue (mod p).

The first few examples of this are equivalent to Euler's conjectures:

$$\left[\frac{2}{p}\right]_3 = 1 \quad \text{if and only if} \quad L \equiv M \equiv 0 \pmod{2}$$

$$\left[\frac{3}{p}\right]_3 = 1 \quad \text{if and only if} \quad M \equiv 0 \pmod{3}$$

$$\left[\frac{5}{p}\right]_3 = 1 \quad \text{if and only if} \quad LM \equiv 0 \pmod{5}$$

$$\left[\frac{7}{p}\right]_3 = 1 \quad \text{if and only if} \quad LM \equiv 0 \pmod{7}$$

$$\left[\frac{11}{p}\right]_3 = 1 \quad \text{if and only if} \quad LM(L-3M)(L+3M) \equiv 0 \pmod{11}$$

$$\left[\frac{13}{p}\right]_3 = 1 \quad \text{if and only if} \quad LM(L-2M)(L+2M) \equiv 0 \pmod{13}$$

Martinet proved

Let p a= q a= 1 (mod 3) be primes, $pq = \frac{1}{4}\left(L^2 + 27M^2\right)$. Then

$$\left[\frac{L}{p}\right]_3 \left[\frac{L}{q}\right]_3 = 1 \quad \text{if and only if} \quad \left[\frac{q}{p}\right]_3 \left[\frac{p}{q}\right]_3 = 1$$

Sharifi proved

Let $p = 1 + 3x + 9x^2$ be prime. Then

Any divisor of x is a cubic residue (mod p).

Eisenstein Integers

Background: In his second monograph on biquadratic reciprocity, Gauss says:

The theorems on biquadratic residues gleam with the greatest simplicity and genuine beauty only when the field of arithmetic is extended to imaginary numbers, so that without restriction, the numbers of the form $a + bi$ constitute the object of study ... we call such numbers integral complex numbers. [bold in the original]

These numbers are now called the ring of Gaussian integers, denoted by Z[i]. Note that i is a fourth root of 1.

In a footnote he adds

The theory of cubic residues must be based in a similar way on a consideration of numbers of the form $a + bh$ where h is an imaginary root of the equation $h^3 = 1$ ... and similarly the theory of residues of higher powers leads to the introduction of other imaginary quantities.

In his first monograph on cubic reciprocity Eisenstein developed the theory of the numbers built up from a cube root of unity; they are now called the ring of Eisenstein integers. Eisenstein said (paraphrasing) "to investigate the properties of this ring one need only consult Gauss's work on Z[i] and modify the proofs". This is not surprising since both rings are unique factorization domains.

The "other imaginary quantities" needed for the "theory of residues of higher powers" are the rings of integers of the cyclotomic number fields; the Gaussian and Eisenstein integers are the simplest examples of these.

Facts and Terminology

Let $\omega = \dfrac{-1+i\sqrt{3}}{2} = e^{\frac{2\pi i}{3}}$ be a complex cube root of unity. The Eisenstein integers Z[ω] are all numbers of the form $a + b\omega$ where a and b are ordinary integers.

Since $\omega^3 - 1 = (\omega - 1)(\omega^2 + \omega + 1) = 0$ and ω `– 1, we have $\omega^2 = -\omega - 1$ and $\omega = -\omega^2 - 1$. Since $\omega^3 = \omega\omega^2 = \omega\bar{\omega} = 1,\ \ \bar{\omega} = \omega^2$ and $\overline{\omega^2} = \omega$ where the bar denotes complex conjugation. Also, $\omega - \bar{\omega} = i\sqrt{3}.$

If $\lambda = a + b\omega$ and $\mu = c + d\omega$,

$\lambda + \mu = (a + c) + (b + d)\omega$ and

$\lambda\ \mu = ac + (ad + bc)\omega + bd\omega^2 = (ac - bd) + (ad + bc - bd)\omega$.

This shows that Z[ω] is closed under addition and multiplication, making it a ring.

The units are the numbers that divide 1. They are ±1, ±ω, and $\pm\omega^2$. They are similar to 1 and -1 in the ordinary integers, in that they divide every number. The units are the powers of -ω, a sixth (not just a third) root of unity.

Given a number $\lambda = a + b\omega$, its conjugate means its complex conjugate $a + b\omega^2 = (a - b) - b\omega$ (not $a - b\omega$), and its associates are its six unit multiples:

$$
\begin{aligned}
\lambda &= a + b\omega \\
\omega\lambda &= -b + (a - b)\omega \\
\omega^2\lambda &= (b - a) - a\omega \\
-\lambda &= -a - b\omega \\
-\omega\lambda &= b + (b - a)\omega \\
-\omega^2\lambda &= (a - b) + a\omega
\end{aligned}
$$

The norm of $\lambda = a + b\omega$ is the product of λ and its conjugate $\mathrm{N}\lambda = \lambda\bar{\lambda} = a^2 - ab + b^2$. From the definition, if λ and μ are two Eisenstein integers, Nλμ = Nλ Nμ; in other words, the norm is a completely multiplicative function. The norm of zero is zero, the norm of any other number is a positive integer. ε is a unit if and only if Nε = 1. Note that the norm is always a– 0 or a– 1 (mod 3).

Z[ω] is a unique factorization domain. The primes fall into three classes:

- 3 is a special case: $3 = -\omega^2(1 - \omega)^2$. It is the only prime in Z divisible by the square of a prime in Z[ω]. In algebraic number theory, 3 is said to ramify in Z[ω].
- Positive primes in Z a– 2 (mod 3) are also primes in Z[ω]. In algebraic number theory, these primes are said to remain inert in Z[ω].
- Positive primes in Z a– 1 (mod 3) are the product of two conjugate primes in Z[ω]. $p = \mathrm{N}\pi = \mathrm{N}\bar{\pi} = \pi\bar{\pi}$ In algebraic number theory, these primes are said to split in Z[ω].

Thus, inert primes are 2, 5, 11, 17, ... and a factorization of the split primes is

$7 = (3 + \omega) \times (2 - \omega),$

$13 = (4 + \omega) \times (3 - \omega),$

$19 = (3 - 2\omega) \times (5 + 2\omega),$

$31 = (1 + 6\omega) \times (-5 - 6\omega), \ldots$

The associates and conjugate of a prime are also primes.

Note that the norm of an inert prime q is Nq = q^2 a– 1 (mod 3).

In order to state the unique factorization theorem, it is necessary to have a way of distinguishing one of the associates of a number. Eisenstein defines a number to be primary if it is a– 2 (mod 3). It is straightforward to show that if gcd(Nλ, 3) = 1 then exactly one associate of λ is primary. A disadvantage of this definition is that the product of two primary numbers is the negative of a primary.

Most modern authors say that a number is primary if it is coprime to 3 and congruent to an ordinary integer (mod $(1 - \omega)^2$), which is the same as saying it is = ±2 (mod 3). There are two reasons to do this: first, the product of two primaries is a primary, and second, it generalizes to all cyclotomic number fields. Under this definition, if gcd(Nλ, 3) = 1 one of λ, $\omega\lambda$, or $\omega^2\lambda$ is primary. A primary under Eisenstein's definition is primary under the modern one, and if λ is primary under the modern one, either λ or $-\lambda$ is primary under Eisenstein's. Since -1 is a cube, this does not affect the statement of cubic reciprocity, but it does affect the unique factorization theorem. This article uses the modern definition, so

The product of two primary numbers is primary and the conjugate of a primary number is also primary.

The unique factorization theorem for Z[ω] is: if $\lambda \neq 0$, then

$$\lambda = \omega^{\mu} (1-\omega)^{\nu} \pi_1^{\alpha_1} \pi_2^{\alpha_2} \pi_3^{\alpha_3} \ldots$$

where $0 \leq \mu \leq 2$, ν e– 0, each π_i is a primary prime, and each $\alpha_i \geq 1$, and this representation is unique, up to the order of the factors.

The notions of congruence and greatest common divisor are defined the same way in Z[ω] as they are for the ordinary integers Z. Because the units divide all numbers, a congruence (mod λ) is also true modulo any associate of λ, and any associate of a GCD is also a GCD.

Cubic Residue Character

An analogue of Fermat's little theorem is true in Z[ω]: if α is not divisible by a prime π,

$$\alpha^{N\pi-1} \equiv 1 \pmod{\pi}$$

Now assume that Nπ ≠ 3, so that Nπ a– 1 (mod 3).

Then $\alpha^{\frac{N\pi-1}{3}}$ makes sense, and $\alpha^{\frac{N\pi-1}{3}} \equiv \omega^k \pmod{\pi}$ for a unique unit ω^k.

This unit is called the cubic residue character of α (mod π) and is denoted by

$$\left(\frac{\alpha}{\pi}\right)_3 = \omega^k \equiv \alpha^{\frac{N\pi-1}{3}} \pmod{\pi}.$$

It has formal properties similar to those of the Legendre symbol.

The congruence $x^3 \equiv \alpha \pmod{\pi}$ is solvable in Z[ω] if and only if $\left(\frac{\alpha}{\pi}\right)_3 = 1.$

$$\left(\frac{\alpha\beta}{\pi}\right)_3 = \left(\frac{\alpha}{\pi}\right)_3\left(\frac{\beta}{\pi}\right)_3$$

$\overline{\left(\frac{\alpha}{\pi}\right)_3} = \left(\frac{\overline{\alpha}}{\overline{\pi}}\right)_3$ where the bar denotes complex conjugation.

if π and θ are associates, $\left(\frac{\alpha}{\pi}\right)_3 = \left(\frac{\alpha}{\theta}\right)_3$

if α a– β (mod π), $\left(\frac{\alpha}{\pi}\right)_3 = \left(\frac{\beta}{\pi}\right)_3$

The cubic character can be extended multiplicatively to composite numbers (coprime to 3) in the "denominator" in the same way the Legendre symbol is generalized into the Jacobi symbol. Like the Jacobi symbol, if the "denominator" of the cubic character is composite, then if the "numerator" is a cubic residue mod the "denominator" the symbol will equal 1, if the symbol does not equal 1 then the "numerator" is a cubic nonresidue, but the symbol can equal 1 when the "numerator" is a nonresidue:

$$\left(\frac{\alpha}{\lambda}\right)_3 = \left(\frac{\alpha}{\pi_1}\right)_3^{\alpha_1} \left(\frac{\alpha}{\pi_2}\right)_3^{\alpha_2} \cdots \quad \text{where} \quad \lambda = \pi_1^{\alpha_1}\pi_2^{\alpha_2}\pi_3^{\alpha_3}\ldots$$

If a and b are ordinary integers, gcd(a, b) = gcd(b, 3) = 1, then

$$\left(\frac{a}{b}\right)_3 = 1.$$

Statement of the Theorem

Let α and β be primary. Then

$$\left(\frac{\alpha}{\beta}\right)_3 = \left(\frac{\beta}{\alpha}\right)_3.$$

There are supplementary theorems for the units and the prime 1 - ω:

Let α = a + bω be primary, a = 3m + 1 and b = 3n. (If a a– 2 (mod 3) replace α with its associate -α; this will not change the value of the cubic characters.) Then

$$\left(\frac{\omega}{\alpha}\right)_3 = \omega^{\frac{1-a-b}{3}} = \omega^{-m-n}, \quad \left(\frac{1-\omega}{\alpha}\right)_3 = \omega^{\frac{a-1}{3}} = \omega^m, \quad \left(\frac{3}{\alpha}\right)_3 = \omega^{\frac{b}{3}} = \omega^n.$$

Davenport–Schmidt Theorem

In mathematics, specifically the area of Diophantine approximation, the Davenport–Schmidt theorem tells us how well a certain kind of real number can be approximated by another kind. Specifically it tells us that we can get a good approximation to irrational numbers that are not quadratic by using either quadratic irrationals or simply rational numbers. It is named after Harold Davenport and Wolfgang M. Schmidt.

Statement

Given a number α which is either rational or a quadratic irrational, we can find unique integers *x*, *y*, and *z* such that *x*, *y*, and *z* are not all zero, the first non-zero one among them is positive, they are relatively prime, and we have

$$x\alpha^2 + y\alpha + z = 0.$$

If α is a quadratic irrational we can take *x*, *y*, and *z* to be the coefficients of its minimal polynomial. If α is rational we will have $x = 0$. With these integers uniquely determined for each such α we can define the *height* of α to be

$$H(\alpha) = \max\{|x|,|y|,|z|\}.$$

The theorem then says that for any real number ξ which is neither rational nor a quadratic irrational, we can find infinitely many real numbers α which *are* rational or quadratic irrationals and which satisfy

$$|\xi - \alpha| < CH(\alpha)^{-3},$$

where

$$C = \begin{cases} C_0 & \text{if } |\xi| < 1 \\ C_0\xi^2 & \text{if } |\xi| > 1. \end{cases}$$

Here we can take C_0 to be any real number satisfying $C_0 > 160/9$.

While the theorem is related to Roth's theorem, its real use lies in the fact that it is effective, in the sense that the constant *C* can be worked out for any given ξ.

Algebraic Number Field

In mathematics, an algebraic number field (or simply number field) *F* is a finite (and hence algebraic) field extension of the field of rational numbers Q. Thus *F* is a field that contains Q and has finite dimension when considered as a vector space over Q. The study of algebraic number fields, and, more generally, of algebraic extensions of the field of rational numbers, is the central topic of algebraic number theory.

Definition

Prerequisites: The notion of algebraic number field relies on the concept of a field. Fields consists of a set of elements together with two operations, namely addition, and multiplication, and some distributivity assumptions. A prominent example of a field is the field of rational numbers, commonly denoted Q, together with its usual operations of addition etc.

Another notion needed to define algebraic number fields is vector spaces. To the extent needed here, vector spaces can be thought of as consisting of sequences (or tuples)

$$(x_1, x_2, \ldots)$$

whose entries are elements of a fixed field, such as the field Q. Any two such sequences can be added by adding the entries one per one. In addition, any sequence can be multiplied by a single element c of the fixed field. These two operations known as vector addition and scalar multiplication satisfy a number of properties that serve to define vector spaces abstractly. Vector spaces are allowed to be "infinite-dimensional", that is to say that the sequences constituting the vector spaces are of infinite length. If, however, the vector space consists of *finite* sequences

$(x_1, x_2, \ldots, x_n)$,

the vector space is said to be of finite dimension, n.

Definition

An algebraic number field (or simply number field) is a finite degree field extension of the field of rational numbers. Here its dimension as a vector space over Q is simply called its degree.

Examples

- The smallest and most basic number field is the field Q of rational numbers. Many properties of general number fields, such as unique factorization, are modelled after the properties of Q.
- The Gaussian rationals, denoted Q(i) (read as "Q adjoined i"), form the first nontrivial example of a number field. Its elements are expressions of the form

$$a+bi$$

where both a and b are rational numbers and i is the imaginary unit. Such expressions may be added, subtracted, and multiplied according to the usual rules of arithmetic and then simplified using the identity

$$i^2 = -1.$$

Explicitly,

$$(a + bi) + (c + di) = (a + c) + (b + d)i,$$

$$(a + bi)\ (c + di) = (ac \text{ “ } bd) + (ad + bc)i.$$

Non-zero Gaussian rational numbers are invertible, which can be seen from the identity

$$(a+bi)\left(\frac{a}{a^2+b^2} - \frac{b}{a^2+b^2}i\right) = \frac{(a+bi)(a-bi)}{a^2+b^2} = 1.$$

It follows that the Gaussian rationals form a number field which is two-dimensional as a vector space over Q.

- More generally, for any square-free integer d, the quadratic field

$$Q(\sqrt{d})$$

is a number field obtained by adjoining the square root of d to the field of rational numbers. Arithmetic operations in this field are defined in analogy with the case of gaussian rational numbers, $d = -1$.

- Cyclotomic field

$$Q(\zeta_n),\ \zeta_n = \exp\ (2\pi i\ /\ n)$$

is a number field obtained from Q by adjoining a primitive nth root of unity ζ_n. This field contains all complex nth roots of unity and its dimension over Q is equal to $\varphi(n)$, where φ is the Euler totient function.

- The real numbers, R, and the complex numbers, C, are fields which have infinite dimension as Q-vector spaces, hence, they are *not* number fields. This follows from the uncountability of R and C as sets, whereas every number field is necessarily countable.
- The set Q^2 of ordered pairs of rational numbers, with the entrywise addition and multiplication is a two-dimensional

commutative algebra over Q. However, it is not a field, since it has zero divisors:

$$(1, 0) \cdot (0, 1) = (1 \cdot 0, 0 \cdot 1) = (0, 0).$$

Algebraicity and Ring of Integers

Generally, in abstract algebra, a field extension F / E is algebraic if every element f of the bigger field F is the zero of a polynomial with coefficients $e_0, ..., e_m$ in E:

$$p(f) = e_m f^m + e_{m-1} f^{m-1} + ... + e_1 f + e_0 = 0.$$

It is a fact that every finite field extension is algebraic (proof: for x in F simply consider x, x^2, x^3 ...we get a linear dependence, i.e. a polynomial x is a root of!). In particular this applies to algebraic number fields, so any element f of an algebraic number field F can be written as a zero of a polynomial with rational coefficients. Therefore, elements of F are also referred to as *algebraic numbers*. Given a polynomial p such that $p(f) = 0$, it can be arranged such that the leading coefficient e_m is one, by dividing all coefficients by it, if necessary. A polynomial with this property is known as a monic polynomial. In general it will have rational coefficients. If, however, its coefficients are actually all integers, f is called an *algebraic integer*. Any (usual) integer z - Z is an algebraic integer, as it is the zero of the linear monic polynomial:

$$p(t) = t - z.$$

It can be shown that any algebraic integer that is also a rational number must actually be an integer, whence the name "algebraic integer". Again using abstract algebra, specifically the notion of a finitely generated module, it can be shown that the sum and the product of any two algebraic integers is still an algebraic integer, it follows that the algebraic integers in F form a ring denoted O_F called the ring of integers of F. It is a subring of (that is, a ring contained in) F. A field contains no zero divisors and this property is inherited by any subring. Therefore, the ring of integers of F is an integral domain. The field F is the field of fractions of the integral domain O_F. This way one can get back and forth between the algebraic number field F and its ring of integers O_F. Rings of algebraic integers have three distinctive properties: firstly, O_F is an integral domain that is integrally closed in its field of fractions

F. Secondly, O_F is a Noetherian ring. Finally, every nonzero prime ideal of O_F is maximal or, equivalently, the Krull dimension of this ring is one. An abstract commutative ring with these three properties is called a *Dedekind ring* (or *Dedekind domain*), in honor of Richard Dedekind, who undertook a deep study of rings of algebraic integers.

Unique Factorization and Class Number

For general Dedekind rings, in particular rings of integers, there is a unique factorization of ideals into a product of prime ideals. However, unlike Z as the ring of integers of Q, the ring of integers of a proper extension of Q need not admit unique factorization of numbers into a product of prime numbers or, more precisely, prime elements. This happens already for quadratic integers, for example in $O_{Q(\sqrt{-5})} = Z[\sqrt{-5}]$, the uniqueness of the factorization fails:

$$6 = 2 \cdot 3 = (1 + \sqrt{-5}) \cdot (1 - \sqrt{-5}),$$

–using the norm it can be shown that these two factorization are actually inequivalent in the sense that the factors do not just differ by a unit in $O_{Q(\sqrt{-5})}$. Euclidean domains are unique factorization domains; for example $Z[i]$, the ring of Gaussian integers, and $Z[\omega]$, the ring of Eisenstein integers, where ω is a third root of unity (unequal to 1), have this property.

ζ-functions, L-functions and Class Number Formula

The failure of unique factorization is measured by the class number, commonly denoted h, the cardinality of the so-called ideal class group. This group is always finite. The ring of integers O_F possesses unique factorization if and only if it is a principal ring or, equivalently, if F has class number 1. Given a number field, the class number is often difficult to compute. The class number problem, going back to Gauss, is concerned with the existence of imaginary quadratic number fields (i.e., $Q(\sqrt{-d})$, $d \geq 1$) with prescribed class number. The class number formula relates h to other fundamental invariants of F. It involves the Dedekind zeta function $\zeta_F(s)$, a function in a complex variable s, defined by

$$\zeta_F(s) := \prod_{\mathfrak{p}} \frac{1}{1 - N(\mathfrak{p})^{-s}}.$$

(The product is over all prime ideals of O_F, $N(\mathfrak{p})$ denotes the norm of the prime ideal or, equivalently, the (finite) number of elements in the residue field $O_F / \mathfrak{p}$. The infinite product converges only for Re(s) > 1, in general analytic continuation and the functional equation for the zeta-function are needed to define the function for all s). The Dedekind zeta-function generalizes the Riemann zeta-function in that $\zeta_Q(s) = \zeta(s)$.

The class number formula states that $\zeta_F(s)$ has a simple pole at $s = 1$ and at this point (its meromorphic continuation to the whole complex plane) the residue is given by

$$\frac{2^{r_1} \cdot (2\pi)^{r_2} \cdot h \cdot \text{Reg}}{w \cdot \sqrt{|D|}}.$$

Here r_1 and r_2 classically denote the number of real embeddings and pairs of complex embeddings of F, respectively. Moreover, Reg is the regulator of F, w the number of roots of unity in F and D is the discriminant of F.

Dirichlet L-functions $L(\chi, s)$ are a more refined variant of $\zeta(s)$. Both types of functions encode the arithmetic behaviour of Q and F, respectively. For example, Dirichlet's theorem asserts that in any arithmetic progression

$$a,\ a + m,\ a + 2m,\ \ldots$$

with coprime a and m, there are infinitely many prime numbers. This theorem is implied by the fact that the Dirichlet L-function is nonzero at $s = 1$. Using much more advanced techniques including algebraic K-theory and Tamagawa measures, modern number theory deals with a description, if largely conjectural, of values of more general L-functions.

Bases for Number Fields

Integral basis: An *integral basis* for a number field F of degree n is a set

$$B = \{b_1, \ldots, b_n\}$$

of n algebraic integers in F such that every element of the ring of integers O_F of F can be written uniquely as a Z-linear combination of elements of B; that is, for any x in O_F we have

$$x = m_1 b_1 + \dots + m_n b_n,$$

where the m_i are (ordinary) integers. It is then also the case that any element of F can be written uniquely as

$$m_1 b_1 + \dots + m_n b_n,$$

where now the m_i are rational numbers. The algebraic integers of F are then precisely those elements of F where the m_i are all integers.

Working locally and using tools such as the Frobenius map, it is always possible to explicitly compute such a basis, and it is now standard for computer algebra systems to have built-in programmes to do this.

Power basis

Let F be a number field of degree n. Among all possible bases of F, there are particular ones known as power bases, that are bases of the form

$$B_x = \{1, x, x^2, \dots, x^{n-1}\}$$

for some element x - F. By the primitive element theorem, for there exists such an x, called a primitive element. If x can be chosen in O_F and such that B_x is a basis of O_F as a free Z-module, then B_x is called a power integral basis, and the field F is called a monogenic field. An example of a number field that is not monogenic was first given by Dedekind. His example is the field obtained by adjoining a root of the polynomial $x^3 - x^2 - 2x - 8$.

Regular Representation, Trace and Determinant

Using the multiplication in F, the elements of the field F may be represented by n-by-n matrices

$$A = A(x) = (a_{ij})_{1 \le i,\, j \le n},$$

by requiring

$$xe_i = \sum_{j=1}^{n} a_{ij} e_j, \quad a_{ij} \in \mathbf{Q}.$$

Here $e_1, \dots, e_n$ is a fixed basis for F, viewed as a Q-vector space. The rational numbers a_{ij} are uniquely determined by x and the choice of a basis since any element of F can be uniquely represented as a linear combination of the basis elements. This way of associating a matrix to any element of the field F is called the *regular representation*. The square matrix A represents the effect of multiplication by x in the given basis. It follows that if the element y of F is represented by a matrix B, then the product xy is represented by the matrix product AB.

Invariants of matrices, such as the trace, determinant, and characteristic polynomial, depend solely on the field element x and not on the basis. In particular, the trace of the matrix $A(x)$ is called the *trace* of the field element x and denoted $\mathrm{Tr}(x)$, and the determinant is called the *norm* of x and denoted $\mathrm{N}(x)$.

By definition, standard properties of traces and determinants of matrices carry over to Tr and N: $\mathrm{Tr}(x)$ is a linear function of x, as expressed by $\mathrm{Tr}(x + y) = \mathrm{Tr}(x) + \mathrm{Tr}(y)$, $\mathrm{Tr}(\lambda x) = \lambda\ \mathrm{Tr}(x)$, and the norm is a multiplicative homogeneous function of degree n: $\mathrm{N}(xy) = \mathrm{N}(x)\ \mathrm{N}(y)$, $\mathrm{N}(\lambda x) = \lambda^n\ \mathrm{N}(x)$. Here λ is a rational number, and x, y are any two elements of F.

The *trace form* derives is a bilinear form defined by means of the trace, as $\mathrm{Tr}(x\ y)$. The *integral trace form*, an integer-valued symmetric matrix is defined as $t_{ij} = \mathrm{Tr}(b_i b_j)$, where $b_1, \ldots, b_n$ is an integral basis for F. The *discriminant* of F is defined as $\det(t)$. It is an integer, and is an invariant property of the field F, not depending on the choice of integral basis.

The matrix associated to an element x of F can also be used to give other, equivalent descriptions of algebraic integers. An element x of F is an algebraic integer if and only if the characteristic polynomial p_A of the matrix A associated to x is a monic polynomial with integer coefficients.

Suppose that the matrix A that represents an element x has integer entries in some basis e. By the Cayley–Hamilton theorem, $p_A(A) = 0$, and it follows that $p_A(x) = 0$, so that x is an algebraic integer. Conversely, if x is an element of F which is a root of a monic polynomial with integer coefficients then the same property holds for the corresponding matrix A.

In this case it can be proven that A is an integer matrix in a suitable basis of F. Note that the property of being an algebraic integer is *defined* in a way that is independent of a choice of a basis in F.

Example

Consider $F = Q(x)$, where x satisfies $x^3 - 11x^2 + x + 1 = 0$. Then an integral basis is $[1, x, 1/2(x^2 + 1)]$, and the corresponding integral trace form is

$$\begin{bmatrix} 3 & 11 & 61 \\ 11 & 119 & 653 \\ 61 & 653 & 3589 \end{bmatrix}.$$

The "3" in the upper left hand corner of this matrix is the trace of the matrix of the map defined by the first basis element (1) in the regular representation of F on F. This basis element induces the identity map on the 3-dimensional vector space, F. The trace of the matrix of the identity map on a 3-dimensional vector space is 3.

The determinant of this is $1304 = 2^3\,163$, the field discriminant; in comparison the root discriminant, or discriminant of the polynomial, is $5216 = 2^5\,163$.

Places

Mathematicians of the nineteenth century assumed that algebraic numbers were a type of complex number. This situation changed with the discovery of p-adic numbers by Hensel in 1897; and now it is standard to consider all of the various possible embeddings of a number field F into its various topological completions at once.

A *place* of a number field F is an equivalence class of absolute values on F. Essentially, an absolute value is a notion to measure the size of elements f of F. Two such absolute values are considered equivalent if they give rise to the same notion of smallness (or proximity). In general, they fall into three regimes. Firstly (and mostly irrelevant), the trivial absolute value $|\ |_0$, which takes the value 1 on all non-zero f in F.

The second and third classes are Archimedean places and non-Archimedean (or ultrametric) places. The completion of F with respect to a place is given in both cases by taking Cauchy sequences in F and dividing out null sequences, that is, sequences $(x_n)_{n - N}$ such that $|x_n|$ tends to zero when n tends to infinity. This can be shown to be a field again, the so-called completion of F at the given place.

For $F = Q$, the following non-trivial norms occur (Ostrowski's theorem): the (usual) absolute value, which gives rise to the complete topological field of the real numbers R. On the other hand, for any prime number p, the p-adic absolute values is defined by

$|q|_p = p^{-n}$, where $q = p^n\, a/b$ and a and b are integers not divisible by p.

In contrast to the usual absolute value, the p-adic norm gets *smaller* when q is multiplied by p, leading to quite different behaviour of Q_p vis-à-vis R.

Archimedean Places

Calculating the archimedean places of F is done as follows: let x be a primitive element of F, with minimal polynomial (over Q) f. Over R, f will generally no longer be irreducible, but its irreducible (real) factors are either of degree one or two. Since there are no repeated roots, there are no repeated factors. The roots r of factors of degree one are necessarily real, and replacing x by r gives an embedding of F into R; the number of such embeddings is equal to the number of real roots of f. Restricting the standard absolute value on R to F gives an archimedean absolute value on F; such an absolute value is also referred to as a *real place* of F. On the other hand, the roots of factors of degree two are pairs of conjugate complex numbers, which allows for two conjugate embeddings into C. Either one of this pair of embeddings can be used to define an absolute value on F, which is the same for both embeddings since they are conjugate. This absolute value is called a *complex place* of F.

If all roots of f above are real or, equivalently, any embedding F ,– C is actually inside R, F is called totally real.

Nonarchimedian or Ultrametric Places

To find the nonarchimedean places, let again f and x be as above. In Q_p, f splits in factors of various degrees, none of which are repeated, and the degrees of which add up to n, the degree of f. For each of these p-adically irreducible factors t, we may suppose that x satisfies t and obtain an embedding of F into an algebraic extension of finite degree over Q_p. Such a local field behaves in many ways like a number field, and the p-adic numbers may similarly play the role of the rationals; in particular, we can define the norm and trace in exactly the same way, now giving functions mapping to Q_p. By using this p-adic norm map N_t for the place t, we may define an absolute value corresponding to a given p-adically irreducible factor t of degree m by $|\theta|_t = |N_t(\theta)|_p^{1/m}$. Such an absolute value is called an ultrametric, non-Archimedean or p-adic place of F.

For any ultrametric place v we have that $|x|_v \leq 1$ for any x in O_F, since the minimal polynomial for x has integer factors, and hence its p-adic factorization has factors in Z_p. Consequently, the norm term (constant term) for each factor is a p-adic integer, and one of these is the integer used for defining the absolute value for v.

Prime Ideals in O_F

For an ultrametric place v, the subset of O_F defined by $|x|_v < 1$ is an ideal P of O_F. This relies on the ultrametricity of v: given x and y in P, then

$$|x + y|_v \leq \max(|x|_v, |y|_v) < 1.$$

Actually, P is even a prime ideal.

Conversely, given a prime ideal P of O_F, a discrete valuation can be defined by setting $v_p(x) = n$ where n is the biggest integer such that $x - P^n$, the n-fold power of the ideal. This valuation can be turned into a ultrametric place. Under this correspondence, (equivalence classes) of ultrametric places of F correspond to prime ideals of O_F. For F = Q, this gives back Ostrowski's theorem: any prime ideal in Z (which is necessarily by a single prime number) corresponds to an non-archimedean place and vice versa. However, for more general number fields, the situation becomes more involved, as will be explained below.

Yet another, equivalent way of describing ultrametric places is by means of localizations of O_F. Given an ultrametric place v on a number field F, the corresponding localization is the subring T of F of all elements x such that $|x|_v \leq 1$. By the ultrametric property T is a ring. Moreover, it contains O_F. For every element

x of F, at least one of x or x^{-1} is contained in T. Actually, since $F^{\times}/T^{\times}$ can be shown to be isomorphic to the integers, T is a discrete valuation ring, in particular a local ring. Actually, T is just the localization of O_F at the prime ideal P. Conversely, P is the maximal ideal of T.

Altogether, there is a three-way equivalence between ultrametric absolute values, prime ideals, and localizations on a number field.

Ramification

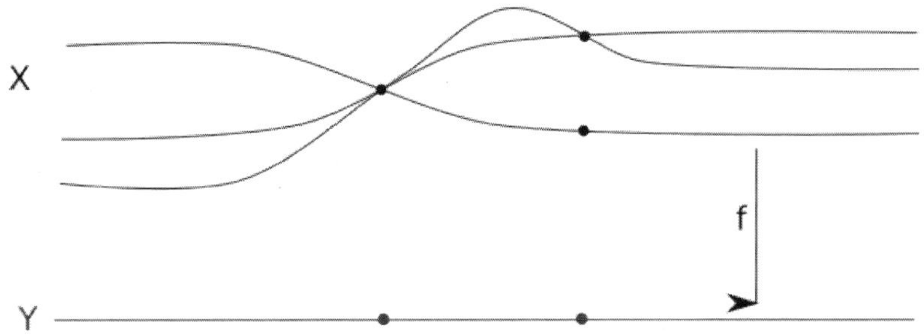

Figure: *Schematic depiction of ramification: the fibres of almost all points in* Y *below consist of three points, except for two points in* Y *marked with dots, where the fibres consist of one and two points (marked in black), respectively. The map* f *is said to be ramified in these points of* Y.

Ramification, generally speaking, describes a geometric phenomenon that can occur with finite-to-one maps (that is, maps $f: X \to Y$ such that the preimages of all points y in Y consist only of finitely many points): the cardinality of the fibres $f^{-1}(y)$ will generally have the same number of points, but it occurs that, in special points y, this number drops. For example, the map

$$\mathbf{C} \to \mathbf{C}, z \mapsto z^n$$

has n points in each fibre over t, namely the n (complex) roots of t, except in $t = 0$, where the fibre consists of only one element, $z = 0$. One says that the map is "ramified" in zero. This is an example of a branched covering of Riemann surfaces. This intuition also serves to define ramification in algebraic number theory. Given a (necessarily finite) extension of number fields F / E, a prime ideal p of O_E generates the ideal pO_F of O_F. This ideal may or may not be a prime ideal, but, according to the Lasker–Noether theorem, always is given by

$$pO_F = q_1^{e_1} q_2^{e_2} \cdots q_m^{e_m}$$

with uniquely determined prime ideals q_i of O_F and numbers (called ramification indices) e_i. Whenever one ramification index is bigger than one, the prime p is said to ramify in F.

The connection between this definition and the geometric situation is delivered by the map of spectra of rings Spec $O_F \rightarrow$ Spec O_E. In fact, unramified morphisms of schemes in algebraic geometry are a direct generalization of unramified extensions of number fields.

Ramification is a purely local property, i.e., depends only on the completions around the primes p and q_i. The inertia group measures the difference between the local Galois groups at some place and the Galois groups of the involved finite residue fields.

An Example

The following example illustrates the notions introduced above. In order to compute the ramification index of Q(x), where

$$f(x) = x^3 - x - 1 = 0,$$

at 23, it suffices to consider the field extension $Q_{23}(x) / Q_{23}$. Up to $529 = 23^2$ (i.e., modulo 529) f can be factored as

$$f(x) = (x + 181)(x^2 - 181x - 38) = gh.$$

Substituting $x = y + 10$ in the first factor g modulo 529 yields $y + 191$, so the valuation $|y|_g$ for y given by g is $|-191|_{23} = 1$. On the other hand the same substitution in h yields y^2 e $161y$ e 161 modulo 529. Since $161 = 7 \times 23$,

$$|y|_h = \sqrt{|161|_{23}} = 1 / \sqrt{23}.$$

Since possible values for the absolute value of the place defined by the factor h are not confined to integer powers of 23, but instead are integer powers of the square root of 23, the ramification index of the field extension at 23 is two.

The valuations of any element of F can be computed in this way using resultants. If, for example $y = x^2 - x - 1$, using the resultant to eliminate x between this relationship and $f = x^3 - x - 1 = 0$ gives $y^3 - 5y^2 + 4y - 1 = 0$. If instead we eliminate with respect to the factors g and h of f, we obtain the corresponding factors for the polynomial for y, and then the 23-adic valuation applied to the constant (norm) term allows us to compute the valuations of y for g and h (which are both 1 in this instance.)

Dedekind Discriminant Theorem

Much of the significance of the discriminant lies in the fact that ramified ultrametric places are all places obtained from factorizations in Q_p where p divides the discriminant. This is even true of the polynomial discriminant; however the converse is also true, that if a prime p divides the discriminant, then there is a p-place which ramifies. For this converse the field discriminant is needed. This is the Dedekind discriminant theorem. In the example above, the discriminant of the number field Q(x) with $x^3 - x - 1 = 0$ is "23, and as we have seen the 23-adic place ramifies. The Dedekind discriminant tells us it is the only ultrametric place which does. The other ramified place comes from the absolute value on the complex embedding of F.

Galois Groups and Galois Cohomology

Generally in abstract algebra, field extensions F / E can be studied by examining the Galois group Gal(F / E), consisting of field automorphisms of F leaving E elementwise fixed. As an example, the Galois group Gal (Q(ζ_n) / Q) of the cyclotomic field extension of degree n is given by $(Z/nZ)^\times$, the group of invertible elements in Z/nZ. This is the first stepstone into Iwasawa theory.

In order to include all possible extensions having certain properties, the Galois group concept is commonly applied to the (infinite) field extension F / F of the algebraic closure, leading to the absolute Galois group G := Gal(F / F) or just Gal(F), and to the extension F / Q. The fundamental theorem of Galois theory links fields in between F and its algebraic closure and closed subgroups of Gal (F). For example, the abelianization (the biggest abelian quotient) G^{ab} of G corresponds to a field referred to as the maximal abelian extension F^{ab} (called so since any further extension is not abelian, i.e., does not have an abelian Galois group). By the Kronecker–Weber theorem, the maximal abelian extension of Q is the extension generated by all roots of unity. For more general number fields, class field theory, specifically the Artin reciprocity law gives an answer by describing G^{ab} in terms of the idele class group. Also notable is the Hilbert class field, the maximal abelian unramified field extension of *F*. It can be shown to be finite over *F*, its Galois group over *F* is isomorphic to the class group of *F*, in particular its degree equals the class number *h* of *F*.

In certain situations, the Galois group acts on other mathematical objects, for example a group. Such a group is then also referred to as a Galois module. This enables the use of group cohomology for the Galois group Gal(*F*), also known as Galois cohomology, which in the first place measures the failure of exactness of taking Gal(*F*)-invariants, but offers deeper insights (and questions) as well. For example, the Galois group *G* of a field extension *L* / *F* acts on $L^\times$, the nonzero elements of *L*. This Galois module plays a significant role in many arithmetic dualities, such as Poitou-Tate duality. The Brauer group of *F*, originally conceived to classify division algebras over *F*, can be recast as a cohomology group, namely $H^2(\text{Gal}\,(F), F^\times)$.

Local-global Principle

Generally speaking, the term "local to global" refers to the idea that a global problem is first done at a local level, which tends to simplify the questions. Then, of course, the information gained in the local analysis has to be put together to get back to some global statement. For example, the notion of sheaves reifies that idea in topology and geometry.

Local and Global Fields

Number fields share a great deal of similarity with another class of fields much used in algebraic geometry known as function fields of algebraic curves over finite fields. An example is $F_p(T)$. They are similar in many respects, for example in that number rings are one-dimensional regular rings, as are the coordinate rings (the quotient fields of which is the function field in question) of curves. Therefore, both types of field are called global fields. In accordance with the philosophy laid out above, they can be studied at a local level first, that is to say, by looking at the corresponding local fields. For number fields *F*, the local fields are the completions of *F* at all places, including the archimedean ones. For function fields, the local fields are completions of the local rings at all points of the curve for function fields.

Many results valid for function fields also hold, at least if reformulated properly, for number fields. However, the study of number fields often poses difficulties and phenomena not encountered in function fields. For example, in function fields, there is no dichotomy into non-archimedean and archimedean places. Nonetheless, function fields often serves as a source of intuition what should be expected in the number field case.

Hasse Principle

A prototypical question, posed at a global level, is whether some polynomial equation has a solution in F. If this is the case, this solution is also a solution in all completions. The local-global principle or Hasse principle asserts that for quadratic equations, the converse holds, as well. Thereby, checking whether such an equation has a solution can be done on all the completions of F, which is often easier, since analytic methods (classical analytic tools such as intermediate value theorem at the archimedean places and p-adic analysis at the nonarchimedean places) can be used. This implication does not hold, however, for more general types of equations. However, the idea of passing from local data to global ones proves fruitful in class field theory, for example, where local class field theory is used to obtain global insights mentioned above. This is also related to the fact that the Galois groups of the completions F_v can be explicitly determined, whereas the Galois groups of global fields, even of Q are far less understood.

Adeles and Ideles

In order to assemble local data pertaining to all local fields attached to F, the adele ring is set up. A multiplicative variant is referred to as ideles.

Dirichlet's Approximation Theorem

In number theory, Dirichlet's theorem on Diophantine approximation, also called Dirichlet's approximation theorem, states that for any real number α and any positive integer N, there exists integers p and q such that $1 \leq q \leq N$ and

$$|q\alpha - p| \leq \frac{1}{N+1}$$

This is a foundational result in diophantine approximation, showing that any real number has a sequence of good rational approximations: in fact an immediate consequence is that for a given irrational α, the inequality

$$\left|\alpha - \frac{p}{q}\right| < \frac{1}{q^2}$$

is satisfied by infinitely many integers p and q. This corollary also shows that the Thue–Siegel–Roth theorem, a result in the other direction, provides essentially the tightest possible bound, in the sense that the limits on rational approximation of algebraic numbers cannot be improved by lowering the exponent $2 + \varepsilon$ beyond 2.

Method of Proof

This theorem is a consequence of the pigeonhole principle. Dirichlet who proved the result used the same principle in other contexts (for example, the Pell equation) and by naming the principle (in German) popularized its use, though its status in textbook terms comes later. The method extends to simultaneous approximation.

Euclidean Division

In mathematics, and more particularly in arithmetic, the Euclidean division is the usual process of division of integers producing a quotient and a remainder. It can be specified precisely by a theorem stating that these exist uniquely with given properties. An integer division algorithm is any effective method for realizing this division. Using the decimal notation of integers (or any other positional notation), long division provides a fairly efficient division algorithm, and other algorithms exist as well. The integer division algorithm is an important ingredient for other algorithms, such as the Euclidean algorithm for finding the greatest common divisor of two integers.

The integral domains to which the Euclidean division may be generalized with similar properties are called Euclidean domains. They include polynomial rings in one variable over a field and Gaussian integers.

Statement of the Theorem

Given two integers a and b, with $b \neq 0$, there exist unique integers q and r such that $a = bq + r$ and 0 de $r < |b|$, where $|b|$ denotes the absolute value of b.

The four integers that appear in this theorem have been given a name: a is called the dividend, b is called the divisor, q is called the quotient and r is called the remainder.

The computation of the quotient and the remainder from the dividend and the divisor is called division or, in case of ambiguity, Euclidean division.

Examples

- If $a = 7$ and $b = 3$, then $q = 2$ and $r = 1$, since $7 = 3 \times 2 + 1$.
- If $a = 7$ and $b = -3$, then $q = -2$ and $r = 1$, since $7 = -3 \times (-2) + 1$.
- If $a = -7$ and $b = 3$, then $q = -3$ and $r = 2$, since $-7 = 3 \times (-3) + 2$.
- If $a = -7$ and $b = -3$, then $q = 3$ and $r = 2$, since $-7 = -3 \times 3 + 2$.

Proof

The proof consists of two parts — first, the proof of the existence of q and r, and second, the proof of the uniqueness of q and r.

Existence

Consider first the case $b < 0$. Setting $b' = -b$ and $q' = -q$, the equation $a = bq + r$ may be rewritten $a = b'q' + r$ and the inequality $0 < r < |b|$ may be rewritten $0 < r < |b'|$. This reduces the existence for the case $b < 0$ to that of the case $b > 0$.

Similarly, if $a < 0$ and $b > 0$, setting $a' = -a$, $q' = -q - 1$ and $r' = b - r$, the equation $a = bq + r$ may be rewritten $a' = bq' + r'$ and the inequality $0 < r < b$ may be rewritten $0 < r' < b$. Thus the proof of the existence is reduced to the case a ee 0 and $b > 0$ and we consider only this case in the remainder of the proof.

Let q_1 and r_1, both nonnegative, such that $a = bq_1 + r_1$, for example $q_1 = 0$ and $r_1 = a$. If $r_1 < b$, we are done. Otherwise $q_2 = q_1 + 1$ and $r_2 = r_1 - b$ satisfy $a = bq_2 + r_2$ and $0 < r_2 < r_1$. Repeating this process one gets eventually $q = q_k$ and $r = r_k$ such that $a = bq + r$ and $0 < r < b$.

This proves the existence and also gives an algorithm to compute the quotient and the remainder. However this algorithm needs q steps and is thus not efficient.

Uniqueness

Suppose there exists q, q', r, r' with 0 dd r, $r' < |b|$ such that $a = bq + r$ and $a = bq' + r'$. Adding the two inequalities 0 dd $r < |b|$ and d$|b| <$ dr' dd 0 yelds d$|b| < r - r' < |b|$, that is $|r - r'| < |b|$.

Subtracting the two equations yields: $b(q' - q) = (r - r')$. Thus $|b|$ divides $|r - r'|$. If $|r - r'|$ 'd 0 this implies $|b| < |r - r'|$, contradicting previous inequality. Thus, $r = r'$ and $b(q' - q) = 0$. As b 'd 0, this implies $q = q'$, proving uniqueness.

Effectiveness

Usually, a existence proof does not provide an algorithm to compute the existing object, but the above proof provides immediately an algorithm. However this is not a very efficient method, as it requires as many steps as the size of the quotient. This is related to the fact that it only uses addition, subtraction and comparison of the integers, without involving multiplication, nor any particular representation of the integers, such as decimal notation.

In terms of decimal notation, long division provides a much more efficient division algorithm. Its generalization to binary notation allows to use it in a computer. However, algorithms that reduce division to multiplication, like Newton–Raphson one, are usually preferred, because they need a time which is proportional to the time of the multiplication needed to verify the result, independently of the multiplication algorithm which is used.

Eisenstein's Theorem

In mathematics, Eisenstein's theorem, named after the German mathematician Gotthold Eisenstein, applies to the coefficients of any power series which is an algebraic function with rational number coefficients. Through the theorem, it is readily demonstrable that a function such as the exponential function must be a transcendental function.

Suppose therefore that

$$\sum a_n t^n$$

is a formal power series with rational coefficients a_n, which has a non-zero radius of convergence in the complex plane, and within it represents an analytic function that is in fact an algebraic function. Let d_n denote the denominator of a_n, as a fraction in lowest terms. Then Eisenstein's theorem states that there is a finite set S of prime numbers p, such that every prime factor of a number d_n is contained in S.

This has an interpretation in terms of p-adic numbers: with an appropriate extension of the idea, the p-adic radius of convergence of the series is at least 1, for almost all p (i.e. the primes outside the finite set S). In fact that statement is a little weaker, in that it disregards any initial partial sum of the series, in a way that may *vary* according to p. For the other primes the radius is non-zero.

Eisenstein's original paper is the short communication *Über eine allgemeine Eigenschaft der Reihen-Entwicklungen aller algebraischen Functionen* (1852), reproduced in Mathematische Gesammelte Werke, Band II, Chelsea Publishing Co., New York, 1975, p. 765–767.

More recently, many authors have investigated precise and effective bounds quantifying the above almost all.

Equidistribution Theorem

In mathematics, the equidistribution theorem is the statement that the sequence

$$a, 2a, 3a, \ldots \bmod 1$$

is uniformly distributed on the circle $\mathbb{R}/\mathbb{Z}$, when a is an irrational number. It is a special case of the ergodic theorem where one takes the normalized angle measure $\mu = \frac{d\theta}{2\pi}$.

History

While this theorem was proved in 1909 and 1910 separately by Hermann Weyl, Wac³aw Sierpiñski and Piers Bohl, variants of this theorem continue to be studied to this day. In 1916, Weyl proved that the sequence $a, 2^2a, 3^2a, \ldots$ mod 1 is uniformly distributed on the unit interval. In 1935, Ivan Vinogradov proved that the sequence $p_n\, a$ mod 1 is uniformly distributed, where p_n is the nth prime. Vinogradov's proof was a byproduct of the odd Goldbach conjecture, that every sufficiently large odd number is the sum of three primes.

George Birkhoff, in 1931, and Aleksandr Khinchin, in 1933, proved that the generalization $x + na$, for almost all x, is equidistributed on any Lebesgue measurable subset of the unit interval. The corresponding generalizations for the Weyl and Vinogradov results were proven by Jean Bourgain in 1988.

Specifically, Khinchin showed that the identity

$$\lim_{n\to\infty}\frac{1}{n}\sum_{k=1}^{n} f((x+ka) \mod 1) = \int_0^1 f(y)dy$$

holds for almost all x and any Lebesgue integrable function f. In modern formulations, it is asked under what conditions the identity

$$\lim_{n\to\infty}\frac{1}{n}\sum_{k=1}^{n} f((x+b_k a) \mod 1) = \int_0^1 f(y)dy$$

might hold, given some general sequence b_k.

One noteworthy result is that the sequence $2^k a$ mod 1 is uniformly distributed for almost all, but not all, irrational a. Similarly, for the sequence $b_k = 2^k$, for every irrational a, and almost all x, there exists a function f for which the sum diverges. In this sense, this sequence is considered to be a universally bad averaging sequence, as opposed to $b_k = k$, which is termed a universally good averaging sequence, because it does not have the latter shortcoming.

A powerful general result is Weyl's criterion, which shows that equidistribution is equivalent to having a non-trivial estimate for the exponential sums formed with the sequence as exponents. For the case of multiples of a, Weyl's criterion reduces the problem to summing finite geometric series.

5

Principles of Number Theorems

Erdõs–Graham Problem

In combinatorial number theory, the Erdõs–Graham problem is the problem of proving that, if the set {2, 3, 4, ...} of integers greater than one is partitioned into finitely many subsets, then one of the subsets can be used to form an Egyptian fraction representation of unity. That is, for every $r > 0$, and every r-colouring of the integers greater than one, there is a finite monochromatic subset S of these integers such that

$$\sum_{n \in S} \frac{1}{n} = 1.$$

In more detail, Paul Erdõs and Ronald Graham conjectured that, for sufficiently large r, the largest member of S could be bounded by b^r for some constant b independent of r. It was known that, for this to be true, b must be at least e.

Ernie Croot proved the conjecture as part of his Ph.D thesis, and later (while a post-doctoral student at UC Berkeley) published the proof in the *Annals of Mathematics*. The value Croot gives for b is very large: it is at most e^{167000}. Croot's result follows as a corollary of a more general theorem stating the existence of Egyptian fraction representations of unity for sets C of smooth numbers in intervals of the form $[X, X^{1+\delta}]$, where C contains sufficiently many numbers so that the sum of their reciprocals is at least six. The Erdõs–Graham conjecture follows from this result by showing that one can find an interval of this form in which the sum of the reciprocals of all smooth numbers is at least $6r$; therefore, if the integers are r-coloured there must be a monochromatic subset C satisfying the conditions of Croot's theorem.

Erdõs–Fuchs Theorem

In mathematics, in the area of combinatorial number theory, the Erdõs–Fuchs theorem is a statement about the number of ways that numbers can be represented as a sum of two elements of a given set, stating that the average order of this number cannot be close to being a linear function.

The theorem is named after Paul Erdõs and Wolfgang Heinrich Johannes Fuchs.

Statement

Let A be a subset of the natural numbers and $r(n)$ denote the number of ways that a natural number n can be expressed as the sum of two elements of A (taking order into account). We consider the average

$$R(n) = \frac{r(1) + r(2) + \cdots + r(n)}{n}.$$

The theorem states that

$$R(n) = C + O\left(n^{-3/4-\varepsilon}\right)$$

cannot hold unless $C = 0$.

Erdõs–Szemerédi Theorem

In arithmetic combinatorics, the Erdõs–Szemerédi theorem, proven by Paul Erdõs and Endre Szemerédi in 1983, asserts the existence of positive constants c, such that

$$\max(|A+A|, |A \cdot A|) \geq c|A|^{1+\varepsilon}$$

whenever A is a finite non-empty set of real numbers of cardinality $|A|$, $A + A = \{a + b : a, b \in A\}$ where is the sum-set of A with itself, and $A \cdot A = \{ab : a, b \in A\}$. Note that it is possible for A+A to be of comparable size to A if A is an arithmetic progression, and it is possible for $A \cdot A$ to be of comparable size to A if A is a geometric progression. The Erdõs–Szemerédi theorem can thus be viewed as an assertion that it is not possible for a large set to behave like an arithmetic progression and as a geometric progression simultaneously. It can also be viewed as an assertion that the real line does not contain any set resembling a finite subring or finite subfield; it is the first example of what is now known as the *sum-product phenomenon*, which is now known to hold in a wide variety of rings and fields, including finite fields.

It was conjectured by Erdõs and Szemerédi that one can take arbitrarily close to 1. The best result in this direction currently is by Solymosi, who showed that one can take arbitrarily close to 1/3.

Euler's Theorem

In number theory, Euler's theorem (also known as the Fermat–Euler theorem or Euler's totient theorem) states that if n and a are coprime positive integers, then

$$a^{\varphi(n)} \equiv 1 \pmod{n}$$

where $\varphi(n)$ is Euler's totient function. (The notation is explained in the article Modular arithmetic.) It was first stated and proved by Leonhard Euler in 1736.

There is a converse of Euler's theorem: if the above congruence is true then a and n must be coprime.

The theorem is a generalization of Fermat's little theorem, and is further generalized by Carmichael's theorem.

The theorem may be used to easily reduce large powers modulo n. For example, consider finding the ones place decimal digit of 7^{222}, i.e. 7^{222} (mod 10). Note that 7 and 10 are coprime, and $\varphi(10) = 4$. So Euler's theorem yields 7^4 a– 1 (mod 10), and we get 7^{222} a– $7^{4\times55+2}$ a– $(7^4)^{55}\times7^2$ a– $1^{55}\times7^2$ a– 49 a– 9 (mod 10).

In general, when reducing a power of a modulo n (where a and n are coprime), one needs to work modulo $\varphi(n)$ in the exponent of a:

if x a– y (mod $\varphi(n)$), then a^x a– a^y (mod n).

Euler's theorem also forms the basis of the RSA encryption system: encryption and decryption in this system together amount to exponentiating the original text by $k\varphi(n)+1$ for some positive integer k, so Euler's theorem shows that the decrypted result is the same as the original.

Faltings' Theorem

In number theory, the Mordell conjecture is the conjecture made by Mordell (1922) that a curve of genus greater than 1 over the field Q of rational numbers has only finitely many rational points. The conjecture was later generalized by replacing Q by a finite extension. It was proved by Gerd Faltings (1983), and is now known as Faltings' theorem.

Background

Let C be a non-singular algebraic curve of genus g over Q. Then the set of rational points on C may be determined as follows:

- Case $g = 0$: no points or infinitely many; C is handled as a conic section.
- Case $g = 1$: no points, or C is an elliptic curve and its rational points form a finitely generated abelian group (*Mordell's Theorem*, later generalized to the Mordell–Weil theorem). Moreover Mazur's torsion theorem restricts the structure of the torsion subgroup.
- Case $g > 1$: according to the Mordell conjecture, now Faltings' Theorem, C has only a finite number of rational points.

Proofs

Faltings' original proof used the known reduction to a case of the Tate conjecture, and a number of tools from algebraic geometry, including the theory of Néron models. A very different proof, based on diophantine approximation, was found by Paul Vojta. A more elementary variant of Vojta's proof was given by Enrico Bombieri.

Consequences

Faltings' 1983 paper had as consequences a number of statements which had previously been conjectured:

- The Mordell conjecture that a curve of genus greater than 1 over a number field has only finitely many rational points;

- The Shafarevich conjecture that there are only finitely many isomorphism classes of abelian varieties of fixed dimension and fixed polarization degree over a fixed number field with good reduction outside a given finite set of places; and
- The *Isogeny theorem* that abelian varieties with isomorphic Tate modules (as Q_l-modules with Galois action) are isogenous.

The reduction of the Mordell conjecture to the Shafarevich conjecture was due to Parshin (1971). A sample application of Faltings' theorem is to a weak form of Fermat's Last Theorem: for any fixed $n > 4$ there are at most finitely many primitive integer solutions to $a^n + b^n = c^n$, since for such n the curve $x^n + y^n = 1$ has genus greater than 1.

Generalizations

Because of the Mordell–Weil theorem, Faltings' theorem can be reformulated as a statement about the intersection of a curve C with a finitely generated subgroup Γ of an abelian variety A. Generalizing by replacing C by an arbitrary subvariety of A and Γ by an arbitrary finite-rank subgroup of A leads to the Mordell–Lang conjecture, which has been proved.

Another higher-dimensional generalization of Faltings' theorem is the Bombieri–Lang conjecture that if X is a pseudo-canonical variety (i.e., variety of general type) over a number field k, then $X(k)$ is not Zariski dense in X. Even more general conjectures have been put forth by Paul Vojta.

The Mordell conjecture for function fields was proved by Manin (1963) and by Grauert (1965). Coleman (1990) found and fixed a gap in Manin's proof.

Fermat Polygonal Number Theorem

In additive number theory, the Fermat polygonal number theorem states that every positive integer is a sum of at most n n-gonal numbers.

That is, every positive number can be written as the sum of three or fewer triangular numbers, and as the sum of four or fewer square numbers, and as the sum of five or fewer pentagonal numbers, and so on. Three such representations of the number 17, for example, arc shown below:

17 = 10 + 6 + 1 (*triangular numbers*)

17 = 16 + 1 (*square numbers*)

17 = 12 + 5 (*pentagonal numbers*).

The theorem is named after Pierre de Fermat, who stated it without proof, promising to write it in a separate work that never appeared. Joseph Louis Lagrange proved the square case in 1770, which states that every positive number can be represented as a sum of four squares, for example, 7 = 4 + 1 + 1 + 1.. Gauss proved the triangular case in 1796, commemorating the occasion by writing in his diary the line "ÅÕÑÇÊÁ! num = Δ + Δ + ≤, and published a proof in his book Disquisitiones Arithmeticae. For this reason, Gauss' result is sometimes known as the Eureka theorem. The full polygonal number theorem was not resolved until it was finally proven by Cauchy in 1813. The proof of Nathanson (1987) is based on the following lemma due to Cauchy:

For odd positive integers a and b such that $b^2 < 4a$ and $3a < b^2 + 2b + 4$ we can find nonnegative integers s, t, u, and v such that $a = s^2 + t^2 + u^2 + v^2$ and $b = s + t + u + v$.

Fermat's Theorem on Sums of Two Squares

In additive number theory, Pierre de Fermat's theorem on sums of two squares states that an odd prime p is expressible as

$$p = x^2 + y^2,$$

with x and y integers, if and only if

$$p \equiv 1 \pmod{4}.$$

For example, the primes 5, 13, 17, 29, 37 and 41 are all congruent to 1 modulo 4, and they can be expressed as sums of two squares in the following ways:

$$5 = 1^2 + 2^2, \quad 13 = 2^2 + 3^2, \quad 17 = 1^2 + 4^2, \quad 29 = 2^2 + 5^2, \quad 37 = 1^2 + 6^2, \quad 41 = 4^2 + 5^2.$$

On the other hand, the primes 3, 7, 11, 19, 23 and 31 are all congruent to 3 modulo 4, and none of them can be expressed as the sum of two squares.

Albert Girard was the first to make the observation (in 1632) and Fermat was first to claim a proof of it. Fermat announced this theorem in a letter to Marin Mersenne dated December 25, 1640; for this reason this theorem is sometimes called *Fermat's Christmas Theorem.*

Since the Brahmagupta–Fibonacci identity implies that the product of two integers that can be written as the sum of two squares is itself expressible as the sum of two squares, by applying Fermat's theorem to the prime factorization of any positive integer n, we see that if all of n's odd prime factors congruent to 3 modulo 4 occur to an even exponent, it is expressible as a sum of two squares. The converse also holds.

Proofs of Fermat's Theorem on Sums of Two Squares

Fermat usually did not write down proofs of his claims, and he did not provide a proof of this statement. The first proof was found by Euler after much effort and is based on infinite descent. He announced it in a letter to Goldbach on April 12, 1749. Lagrange gave a proof in 1775 that was based on his study of quadratic forms. This proof was simplified by Gauss in his *Disquisitiones Arithmeticae* (art. 182). Dedekind gave at least two proofs based on the arithmetic of the Gaussian integers. There is an elegant proof using Minkowski's theorem about convex sets. Simplifying an earlier short proof due to Heath-Brown (who was inspired by Liouville's idea), Zagier presented a one-sentence proof of Fermat's assertion.

Related Results

Fermat announced two related results fourteen years later. In a letter to Blaise Pascal dated September 25, 1654 he announced the following two results for odd primes :

- $p = x^2 + 2y^2 \Leftrightarrow p \equiv 1 \text{ or } p \equiv 3 \pmod{8}$,
- $p = x^2 + 3y^2 \Leftrightarrow p \equiv 1 \pmod{3}$.

He also wrote:

If two primes which end in 3 or 7 and surpass by 3 a multiple of 4 are multiplied, then their product will be composed of a square and the quintuple of another square.

In other words, if p, q are of the form $20k + 3$ or $20k + 7$, then $pq = x^2 + 5y^2$. Euler later extended this to the conjecture that

- $p = x^2 + 5y^2 \Leftrightarrow p \equiv 1 \text{ or } p \equiv 9 \pmod{20}$,
- $2p = x^2 + 5y^2 \Leftrightarrow p \equiv 3 \text{ or } p \equiv 7 \pmod{20}$.

  Both Fermat's assertion and Euler's conjecture were established by Lagrange.

Freiman's Theorem

In mathematics, Freiman's theorem is a combinatorial result in number theory. In a sense it accounts for the approximate structure of sets of integers that contain a high proportion of their internal sums, taken two at a time.

The formal statement is:

Let A be a finite set of integers such that the sumset

$$A + A$$

is small, in the sense that

$$|A + A| < c|A|$$

for some constant . There exists an n-dimensional arithmetic progression of length

$$c'|A|$$

that contains A, and such that c' and n depend only on c.

A simple instructive case is the following. We always have $|A+A|$ e– $2|A|-1$, with equality precisely when A is an arithmetic progression.

This result is due to Gregory Freiman (1966). Much interest in it, and applications, stemmed from a new proof by Imre Z. Ruzsa (1994).

Fundamental Lemma (Langlands Programme)

In the theory of automorphic forms, an area of mathematics, the fundamental lemma relates orbital integrals on a reductive group over a local field to stable orbital integrals on its endoscopic groups. It was conjectured by Langlands (1983) in the course of developing the Langlands programme.

The fundamental lemma was proved by Gérard Laumon and Ngô B£o Chβu in the case of unitary groups and then by Ngô for general reductive groups, building on a series of important reductions made by Jean-Loup Waldspurger to the case of Lie algebras. Time magazine placed Ngô's proof on the list of the "Top 10 scientific discoveries of 2009□ □In 2010 Ngô was awarded the Fields medal for this proof.

Motivation and History

Robert Langlands outlined a strategy for proving local and global Langlands conjectures using the Arthur–Selberg trace formula, but in order for this approach to work, the geometric sides of the trace formula for different groups must be related in a particular way. This relationship takes the form of identities between orbital integrals on reductive groups *G* and *H* over a nonarchimedean local field *F*, where the group *H*, called an endoscopic group of *G*, is constructed from *G* and some additional data.

The first case considered was $G = SL_2$ (Labesse & Langlands 1979). Langlands and Shelstad (1987) then developed the general framework for the theory of endoscopic transfer and formulated specific conjectures. However, during the next two decades only partial progress was made towards proving the fundamental lemma. Harris called it a "bottleneck limiting progress on a host of arithmetic questions". Langlands himself, writing on the origins of endoscopy, commented:

> *... it is not the fundamental lemma as such that is critical for the analytic theory of automorphic forms and for the arithmetic of Shimura varieties; it is the stabilized (or stable) trace formula, the reduction of the trace formula itself to the stable trace formula for a group and its endoscopic groups, and the stabilization of the Grothendieck–Lefschetz formula. None of these are possible without the fundamental lemma and its absence rendered progress almost impossible for more than twenty years.*

Statement

The fundamental lemma states that an orbital integral O for a group G is equal to a stable orbital integral SO for an endoscopic group H, up to a transfer factor Δ (Nadler 2012):

$$SO_{\gamma_H}(1_{K_H}) = \Delta(\gamma_H, \gamma_G) O^{\kappa}_{\gamma_G}(1_{K_G})$$

where

- F is a local field
- G is an unramified group defined over F, in other words a quasi-split reductive group defined over F that splits over an unramified extension of F
- H is an unramified endoscopic group of G associated to κ
- K_G and K_H are hyperspecial maximal compact subgroups of G and H, which means roughly that they are the subgroups of points with coefficients in the ring of integers of F.
- 1_{KG} and 1_{KH} are the characteristic functions of K_G and K_H.
- $\Delta(\tilde{a}_H, \tilde{a}_G)$ is a transfer factor, a certain elementary expression depending on γ_H and γ_G
- γ_H and γ_G are elements of G and H representing stable conjugacy classes, such that the stable conjugacy class of G is the transfer of the stable conjugacy class of H.
- κ is a character of the group of conjugacy classes in the stable conjugacy class of γ_G
- SO and O are stable orbital integrals and orbital integrals depending on their parametres.

Approaches

Shelstad (1982) proved the fundamental lemma for archimedian fields.

Waldspurger (1991) verified the fundamental lemma for general linear groups.

Kottwitz (1992) and Blasius & Rogawski (1992) verified some cases of the fundamental lemma for 3-dimensional unitary groups.

Hales (1997) and Weissauer (2009) verified the fundamental lemma for the symplectic and general symplectic groups Sp_4, GSp_4.

A paper of George Lusztig and David Kazhdan pointed out that orbital integrals could be interpreted as counting points on certain algebraic varieties over finite fields.

Further, the integrals in question can be computed in a way that depends only on the residue field of *F*; and the issue can be reduced to the Lie algebra version of the orbital integrals. Then the problem was restated in terms of the Springer fibre of algebraic groups.

The circle of ideas was connected to a purity conjecture; Laumon gave a conditional proof based on such a conjecture, for unitary groups. Laumon and Ngô (2008) then proved the fundamental lemma for unitary groups, using Hitchin fibration introduced by Ngô (2006), which is an abstract geometric analogue of the Hitchin system of complex algebraic geometry. Waldspurger (2006) showed for Lie algebras that the function field case implies the fundamental lemma over all local fields, and Waldspurger (2008) showed that the fundamental lemma for Lie algebras implies the fundamental lemma for groups. .

Fundamental Lemma of Sieve Theory

In number theory, the fundamental lemma of sieve theory is any of several results that systematize the process of applying sieve methods to particular problems. Halberstam & Richert write:

A curious feature of sieve literature is that while there is frequent use of Brun's *method* there are only a few attempts to formulate a general Brun *theorem* (such as Theorem 2.1); as a result there are surprisingly many papers which repeat in considerable detail the steps of Brun's argument.

Diamond & Halberstam attribute the terminology *Fundamental Lemma* to Jonas Kubilius.

Common Notation

We use these notations:

- A is a set of X positive integers, and A_d is its subset of integers divisible by d
- $w(d)$ and R_d are functions of A and of d that estimate the number of elements of A that are divisible by d, according to the formula

$$|A_d| = \frac{w(d)}{d} X + R_d.$$

Thus $w(d) / d$ represents an approximate density of members divisible by d, and R_d represents an error or remainder term.

- P is a set of primes, and $P(z)$ is its the product of those primes $\leq z$
- $S(A, P, z)$ is the number of elements of A not divisible by any prime in P that is $\leq z$
- κ is a constant, called the sifting density, that appears in the assumptions below. It is a weighted average of the number of residue classes sieved out by each prime.

Fundamental Lemma of the Combinatorial Sieve

This formulation is from Tenenbaum. Other formulations are in Halberstam & Richert, in Greaves, and in Friedlander & Iwaniec. We make the assumptions:

- $w(d)$ is a multiplicative function.
- The sifting density κ satisfies, for some constant C and any real numbers η and ξ with $2 \leq \eta \leq \xi$:

$$\prod_{\eta \leq p \leq \xi} \left(1 - \frac{w(p)}{p}\right)^{-1} < \left(\frac{\ln \xi}{\ln \eta}\right)^{\kappa} \left(1 + \frac{C}{\ln \eta}\right).$$

There is a parametre u e– 1 that is at our disposal. We have uniformly in A, X, z, and u that

$$S(a, P, z) = X \prod_{p \leq z, p \in P} \left(1 - \frac{w(p)}{p}\right) \{1 + O(u^{-u/2})\} + O\left(\sum_{d \leq z^u, d | P(z)} |R_d|\right).$$

In applications we pick u to get the best error term. In the sieve it represents the number of levels of the inclusion–exclusion principle.

Fundamental Lemma of the Selberg Sieve

This formulation is from Halberstam & Richert. Another formulation is in Diamond & Halberstam.

We make the assumptions:

- $w(d)$ is a multiplicative function.

- The sifting density κ satisfies, for some constant C and any real numbers η and ξ with $2 \le \eta \le \xi$:

$$\sum_{\eta \le p \le \xi} \frac{w(p)\ln p}{p} < \kappa \ln\frac{\xi}{\eta} + C.$$

- $w(p) / p < 1 - c$ for some small fixed c and all p
- $| R_d | \le \omega(d)$ where $\omega(d)$ is the number of distinct prime divisors of d.

The fundamental lemma has almost the same form as for the combinatorial sieve. Write $u = \ln X / \ln z$. The conclusion is:

$$S(a,P,z) = X \prod_{p \le z,\, p \in P} \left(1 - \frac{w(p)}{p}\right) \{1 + O(e^{-u/2})\}.$$

Note that u is no longer an independent parametre at our disposal, but is controlled by the choice of z.

Note that the error term here is weaker than for the fundamental lemma of the combinatorial sieve. Halberstam & Richert remark: "Thus it is not true to say, as has been asserted from time to time in the literature, that Selberg's sieve is always better than Brun's."

Gelfond–Schneider Theorem

In mathematics, the Gelfond–Schneider theorem establishes the transcendence of a large class of numbers. It was originally proved independently in 1934 by Aleksandr Gelfond and Theodor Schneider. The Gelfond–Schneider theorem answers affirmatively Hilbert's seventh problem.

Statement

If α and β are algebraic numbers with $\alpha \ne 0,1$ and if β is not a rational number, then any value of $\alpha^\beta = \exp(\beta \log \alpha)$ is a transcendental number.

Comments

- The values of α and β are not restricted to real numbers; complex numbers are allowed (and never rational, even if both the real and imaginary parts are rational).
- In general, $\alpha^\beta = \exp(\beta \log \alpha)$ is multivalued, where "log" stands for the complex logarithm. This accounts for the phrase "any value of" in the theorem's statement.

- An equivalent formulation of the theorem is the following: if α and γ are nonzero algebraic numbers, and we take any non-zero logarithm of α, then (log γ)/(log α) is either rational or transcendental. This may be expressed as saying that if log α, log γ are linearly independent over the rationals, then they are linearly independent over the algebraic numbers. The generalisation of this statement to several logarithms of algebraic numbers is in the domain of transcendence theory.
- If the restriction that α and β be algebraic is removed, the statement does not remain true in general. For example,

$$\left(\sqrt{2}^{\sqrt{2}}\right)^{\sqrt{2}} = \sqrt{2}^{\sqrt{2}\cdot\sqrt{2}} = \sqrt{2}^{2} = 2.$$

Here, α is "2^{-2} which (as proved by the theorem itself) is transcedental rather than algebraic. Similarly, if you put α = 3 and β = (log 2)/(log 3), which is transcendental, then $\alpha^\beta = 2$ is algebraic. A characterization of the values for α and β which yield a transcendental α^β is not known.

- Kurt Mahler proved the *p*-adic analogue of the theorem: if α and β are in C_p, the completion of the algebraic closure of Q_p, and they are algebraic over Q, and if $|\alpha$ " $1|_p < 1$ and $|\beta$ " $1|_p < 1$, then $(\log_p \alpha)/(\log_p \beta)$ is either rational or transcendental, where $\log_p$ is the *p*-adic logarithm function.

Corollaries

The transcendence of the following numbers follows immediately from the theorem:

- The numbers $2^{\sqrt{2}}$ (the Gelfond–Schneider constant) and its square root $\sqrt{2}^{\sqrt{2}}$.
- The number $e^{\pi} = \left(e^{i\pi}\right)^{-i} = (-1)^{-i} = 23.14069263\ldots$ (Gelfond's constant), as well as $i^{i} = \left(e^{i\pi/2}\right)^{i} = e^{-\pi/2} = 0.207879576\ldots$.

Glaisher's Theorem

In number theory, Glaisher's theorem is an identity useful to the study of integer partitions. It is named for James Whitbread

Lee Glaisher. It states that the number of partitions of an integer N into parts not divisible by d is equal to the number of partitions of the form $N = N_1 + \cdots + N_k$ where $N_i \geq N_{i+1}$ and $N_i \geq N_{i+d-1} + 1$, that is, partitions in which no part is repeated d or more times. When $d = 2$ this becomes the special case, known as Euler's theorem, that the number of partitions of into distinct parts is the same as the number of partitions of into odd parts.

Similar Theorems

If instead of counting the number of partitions with distinct parts we count the number of partitions with parts differing by at least 2, a theorem similar to Euler's theorem known as Rogers' theorem (after Leonard James Rogers) is obtained:

The number of partitions whose parts differ by at least 2 is equal to the number of partitions involving only numbers congruent to 1 or 4 (mod 5). For example, there are 6 partitions of 10 into parts differing by at least 2, namely 10, 9+1, 8+2, 7+3, 6+4, 6+3+1; and 6 partitions of 10 involving only 1, 4, 6, 9 ..., namely 9+1, 6+4, 6+1+1+1+1, 4+4+1+1, 4+1+1+1+1+1+1, 1+1+1+1+1+1+1+1+1+1. The theorem was discovered independently by Schur and Ramanujan.

Hasse–Arf Theorem

In mathematics, specifically in local class field theory, the Hasse–Arf theorem is a result concerning jumps of a filtration of the Galois group of a finite Galois extension. A special case of it was originally proved by Helmut Hasse, and the general result was proved by Cahit Arf.

Statement

Higher Ramification Groups

The theorem deals with the upper numbered higher ramification groups of a finite abelian extension L/K. So assume L/K is a finite Galois extension, and that v_K is a discrete normalised valuation of K, whose residue field has characteristic $p > 0$, and which admits a unique extension to L, say w. Denote by v_L the associated normalised valuation ew of L and let be the valuation ring of L under v_L. Let L/K have Galois group G and define the s-th ramification group of L/K for any real $s \geq -1$ by

$$G_s(L/K) = \{\sigma \in G : v_L(\sigma a - a) \geq s + 1 \text{ for all } a \in \mathcal{O}\}.$$

So, for example, G_{-1} is the Galois group G. To pass to the upper numbering one has to define the function $\psi_{L/K}$ which in turn is the inverse of the function $\eta_{L/K}$ defined by

$$\eta_{L/K}(s) = \int_0^s \frac{dx}{|G_0 : G_x|}.$$

The upper numbering of the ramification groups is then defined by $G^t(L/K) = G_s(L/K)$ where $s = \psi_{L/K}(t)$.

These higher ramification groups $G^t(L/K)$ are defined for any real $t \geq -1$, but since v_L is a discrete valuation, the groups will change in discrete jumps and not continuously. Thus we say that t is a jump of the filtration $\{G^t(L/K) : t \geq -1\}$ if $G^t(L/K) \neq G^u(L/K)$ for any $u > t$. The Hasse–Arf theorem tells us the arithmetic nature of these jumps

Statement of the Theorem

With the above set up, the theorem states that the jumps of the filtration $\{G^t(L/K) : t \geq -1\}$ are all rational integers.

Hasse–Minkowski Theorem

The Hasse–Minkowski theorem is a fundamental result in number theory which states that two quadratic forms over a number field are equivalent if and only if they are equivalent *locally at all places,* i.e. equivalent over every completion of the field (which may be real, complex, or p-adic). A special case is that a quadratic space over a number field is isotropic if and only if it is isotropic locally everywhere, or equivalently, that a quadratic form over a number field nontrivially represents zero if and only if this holds for all completions of the field. The theorem was proved in the case of the field of rational numbers by Hermann Minkowski and generalized to number fields by Helmut Hasse. The same statement holds even more generally for all global fields.

The importance of the Hasse–Minkowski theorem lies in the novel paradigm it presented for answering arithmetical questions: in order to determine whether an equation of a certain type has a solution in rational numbers, it is sufficient to test whether it has solutions over complete fields of real and p-adic numbers, where analytic considerations, such as Newton's method and its p-adic analogue, Hensel's lemma, apply. This is encapsulated in the idea of a local-global principle, which is one of the most fundamental techniques in arithmetic geometry.

Application to the Classification of Quadratic Forms

The Hasse–Minkowski theorem reduces the problem of classifying quadratic forms over a number field K up to equivalence to the set of analogous but much simpler questions over local fields.

Basic invariants of a nonsingular quadratic form are its dimension, which is a positive integer, and its discriminant modulo the squares in K, which is an element of the multiplicative group K^*/K^{*2}. In addition, for every place v of K, there is an invariant coming from the completion K_v.

- *Case of R.* By Sylvester's law of inertia, the signature (or, alternatively, the negative index of inertia) is a complete invariant.
- *Case of* C. All nonsingular quadratic forms of the same dimension are equivalent.
- *Case of Q_p and its algebraic extensions.* Forms of the same dimension are classified up to equivalence by their Hasse invariant.

These invariants must satisfy some compatibility conditions: a parity relation (the sign of the discriminant must match the negative index of inertia) and a product formula (a local–global relation). Conversely, for every set of invariants satisfying these relations, there is a quadratic form over K with these invariants.

Hilbert's Irreducibility Theorem

In number theory, Hilbert's irreducibility theorem, conceived by David Hilbert, states that every finite number of irreducible polynomials in a finite number of variables and having rational number coefficients admit a common specialization of a proper subset of the variables to rational numbers such that all the polynomials remain irreducible.

This theorem is a prominent theorem in number theory.

Formulation of the Theorem

Hilbert's irreducibility theorem. Let

$$f_1(X_1,\ldots,X_r,Y_1,\ldots,Y_s),\ldots,f_n(X_1,\ldots,X_r,Y_1,\ldots,Y_s)$$

be irreducible polynomials in the ring

$$\mathbb{Q}[X_1,\ldots,X_r,Y_1,\ldots,Y_s].$$

Then there exists an r-tuple of rational numbers $(a_1,\ldots,a_r)$ such that

$$f_1(a_1,\ldots,a_r,Y_1,\ldots,Y_s),\ldots,f_n(a_1,\ldots,a_r,Y_1,\ldots,Y_s)$$

are irreducible in the ring

$$\mathbb{Q}[Y_1,\ldots,Y_s].$$

Remarks.

- It follows from the theorem that there are infinitely many r-tuples. In fact the set of all irreducible specialization, called Hilbert set, is large in many senses. For example, this set is Zariski dense in $\mathbb{Q}^r$
- There are always (infinitely many) integer specializations, i.e., the assertion of the theorem holds even if we demand $(a_1,\ldots,a_r)$ to be integers.
- There are many Hilbertian fields, i.e., fields satisfying Hilbert's irreducibility theorem. For example, global fields are Hilbertian.
- The irreducible specialization property stated in the theorem is the most general. There are many reductions, e.g., it suffices to take $n = r = s = 1$ in the definition. A recent result of Bary-Soroker shows that for a field K to be Hilbertian it suffices to consider the case of $n = r = s = 1$ and $f = f_1$ absolutely irreducible, that is, irreducible in the ring $K^{\text{alg}}[X,Y]$, where K^{alg} is the algebraic closure of K.

Applications

Hilbert's irreducibility theorem has numerous applications in number theory and algebra. For example:

- The inverse Galois problem, Hilbert's original motivation. The theorem almost immediately implies that if a finite group G can be realized as the Galois group of a Galois extension N of

$$E = \mathbb{Q}(X_1, \ldots, X_r),$$

then it can be specialized to a Galois extension N_0 of the rational numbers with G as its Galois group.

- Construction of elliptic curves with large rank.
- Hilbert's irreducibility theorem is used as a step in the Andrew Wiles proof of Fermat's last theorem.
- If a polynomial $g(x) \in \mathbb{Z}[x]$ is a perfect square for all large integer values of x, then $g(x)$ is the square of a polynomial in $\mathbb{Z}[x]$. This follows from Hilbert's irreducibility theorem with $n = r = s = 1$ and $f_1(X, Y) = Y^2 - g(X)$.

(More elementary proofs exist.) The same result is true when "square" is replaced by "cube", "fourth power", etc.

Generalizations

It has been reformulated and generalized extensively, by using the language of algebraic geometry.

Hurwitz's Theorem (Number Theory)

In number theory, Hurwitz's theorem, named after Adolf Hurwitz, gives a bound on a Diophantine approximation. The theorem states that for every irrational number ξ there are infinitely many rationals m/n such that

$$\left|\xi - \frac{m}{n}\right| < \frac{1}{\sqrt{5}n^2}.$$

The hypothesis that ξ is irrational cannot be omitted. Moreover the constant is the best possible; if we replace by any number and we let (the golden ratio) then there exist only *finitely* many rational numbers m/n such that the formula above holds.

Jacobi's Four-square Theorem

In 1834, Carl Gustav Jakob Jacobi found an exact formula for the total number of ways a given positive integer n can be represented as the sum of four squares. Two representations are considered different if their terms are in different order or if the integer being squared (not just the square) is different; to illustrate, these are three of the eight different ways to represent 1:

$$1^2 + 0^2 + 0^2 + 0^2$$
$$0^2 + 1^2 + 0^2 + 0^2$$
$$(-1)^2 + 0^2 + 0^2 + 0^2.$$

The number of ways to represent n as the sum of four squares is eight times the sum of the divisors of n if n is odd and 24 times the sum of the odd divisors of n if n is even, i.e.

$$r_4(n) = \begin{cases} 8\sum_{m|n} m & \text{if } n \text{ is odd} \\ \text{[}12pt\text{]}24\sum_{\substack{m|n \\ m \text{ odd}}} m & \text{if } n \text{ is even.} \end{cases}$$

Equivalently, it is eight times the sum of all its divisors which are not divisible by 4, i.e.

$$r_4(n) = 8 \sum_{m:4\nmid m|n} m.$$

In particular, for a prime number p we have the explicit formula $r_4(p) = 8(p + 1)$.

The proof shows that the Theta series for the lattice Z^4 is a modular form of a certain level, and hence equals a linear combination of Eisenstein series.

Jurkat–Richert Theorem

The Jurkat–Richert theorem is a mathematical theorem in sieve theory. It is a key ingredient in proofs of Chen's theorem on Goldbach's conjecture. It was proved in 1965 by Wolfgang B. Jurkat and Hans-Egon Richert.

Statement of the Theorem

This formulation is from Diamond & Halberstam. Other formulations are in Jurkat & Richert, Halberstam & Richert, and Nathanson.

Suppose A is a finite sequence of integers and P is a set of primes. Write A_d for the number of items in A that are divisible by d, and write $P(z)$ for the product of the elements in P that are less than z. Write $\omega(d)$ for a multiplicative function such that $\omega(p)/$

p is approximately the proportion of elements of A divisible by p, write X for any convenient approximation to $|A|$, and write the remainder as

$$r_A(d) = |A_d| - \frac{\omega(d)}{d} X.$$

Write $S(A,P,z)$ for the number of items in A that are relatively prime to $P(z)$. Write

$$V(z) = \prod_{p \in P, p<z} \left(1 - \frac{\omega(p)}{p}\right).$$

Write $\nu(m)$ for the number of distinct prime divisors of m. Write F_1 and f_1 for functions satisfying certain difference differential equations.

We assume the dimension (sifting density) is 1: that is, there is a constant C such that for $2 \le z < w$ we have

$$\prod_{z \le p < w} \left(1 - \frac{\omega(p)}{p}\right)^{-1} \le \left(\frac{\log w}{\log z}\right)\left(1 + \frac{C}{\log z}\right).$$

Then the Jurkat–Richert theorem states that for any numbers y and z with $2 \le z \le y \le X$ we have

$$S(A,P,z) \le XV(z)\left(F_1\left(\frac{\log y}{\log z}\right) + O\left(\frac{(\log\log y)^{3/4}}{(\log y)^{1/4}}\right)\right)$$

$$+ \sum_{m \mid P(z), m<y} 4^{\nu(m)} |r_A(m)|$$

and

$$S(A,P,z) \ge XV(z)\left(f_1\left(\frac{\log y}{\log z}\right) - O\left(\frac{(\log\log y)^{3/4}}{(\log y)^{1/4}}\right)\right)$$

$$- \sum_{m \mid P(z), m<y} 4^{\nu(m)} |r_A(m)|.$$

Kaplansky's Theorem on Quadratic Forms

In mathematics, Kaplansky's theorem on quadratic forms is a result on simultaneous representation of primes by quadratic forms.

It was proved in 2003 by Canadian mathematician Irving Kaplansky (1917–2006).

Statement of the Theorem

Kaplansky's theorem states that a prime p congruent to 1 modulo 16 is representable by both or none of $x^2 + 32y^2$ and $x^2 + 64y^2$, whereas a prime p congruent to 9 modulo 16 is representable by exactly one of these quadratic forms.

This is remarkable since the primes represented by each of these forms individually are *not* describable by congruence conditions.

Proof

Kaplansky's proof uses th facts that 2 is a 4th power modulo p if and only if p is representable by $x^2 + 64y^2$, and that "4 is an 8th power modulo p if and only if p is representable by $x^2 + 32y^2$.

Examples

- The prime $p = 17$ is congruent to 1 modulo 16 and is representable by neither $x^2 + 32y^2$ nor $x^2 + 64y^2$.
- The prime p=113 is congruent to 1 modulo 16 and is representable by both $x^2 + 32y^2$ and x^2+$64y^2$ (since $113 = 9^2 + 32\times1^2$ and $113 = 7^2 + 64\times1^2$).
- The prime $p = 41$ is congruent to 9 modulo 16 and is representable by $x^2 + 32y^2$ (since $41 = 3^2 + 32\times1^2$), but not by $x^2 + 64y^2$.
- The prime $p = 73$ is congruent to 9 modulo 16 and is representable by $x^2 + 64y^2$ (since $73 = 3^2 + 64\times1^2$), but not by $x^2 + 32y^2$.

Similar Results

Five results similar to Kaplansky's theorem are known:

- A prime p congruent to 1 modulo 20 is representable by both or none of $x^2 + 20y^2$ and $x^2 + 100y^2$, whereas a prime p congruent to 9 modulo 20 is representable by exactly one of these quadratic forms.
- A prime p congruent to 1, 16 or 22 modulo 39 is representable by both or none of $x^2 + xy + 10y^2$ and $x^2 + xy + 127y^2$, whereas a prime p congruent to 4, 10 or 25

modulo 39 is representable by exactly one of these quadratic forms.

- A prime p congruent to 1, 16, 26, 31 or 36 modulo 55 is representable by both or none of $x^2 + xy + 14y^2$ and $x^2 + xy + 69y^2$, whereas a prime p congruent to 4, 9, 14, 34 or 49 modulo 55 is representable by exactly one of these quadratic forms.
- A prime p congruent to 1, 65 or 81 modulo 112 is representable by both or none of $x^2 + 14y^2$ and $x^2 + 448y^2$, whereas a prime p congruent to 9, 25 or 57 modulo 112 is representable by exactly one of these quadratic forms.
- A prime p congruent to 1 or 169 modulo 240 is representable by both or none of $x^2 + 150y^2$ and $x^2 + 960y^2$, whereas a prime p congruent to 49 or 121 modulo 240 is representable by exactly one of these quadratic forms.

It is conjectured that there are no other similar results involving definite forms.

Kummer's Theorem

In mathematics, Kummer's theorem, named after Ernst Kummer, is as follows. Given integers n e– m e– 0 and a prime number p, the maximum integer k such that p^k divides the binomial coefficient $\binom{n}{m}$ is equal to the number of carries when m is added to $n - m$ in base p.

Lagrange's Four-square Theorem

Lagrange's four-square theorem, also known as Bachet's conjecture, states that any natural number can be represented as the sum of four integer squares

$$p = a_0^2 + a_1^2 + a_2^2 + a_3^2$$

where the four numbers a_0, a_1, a_2, a_3 are integers. For illustration, 3, 31 and 310 can be represented as the sum of four squares as follows:

$$3 = 1^2 + 1^2 + 1^2 + 0^2$$

$$31 = 5^2 + 2^2 + 1^2 + 1^2$$

$$310 = 17^2 + 4^2 + 2^2 + 1^2.$$

This theorem was proven by Joseph Louis Lagrange in 1770, and corresponds to Fermat's theorem on sums of two squares.

Historical development

The theorem appears in the *Arithmetica* of Diophantus, translated into Latin by Bachet in 1621.

Adrien-Marie Legendre improved on the theorem in 1798 by stating that a positive integer can be expressed as the sum of three squares if and only if it is not of the form $4^k(8m + 7)$. His proof was incomplete, leaving a gap which was later filled by Carl Friedrich Gauss.

Proof Using the Hurwitz Integers

One of the ways to prove the theorem relies on Hurwitz quaternions, which are the analog of integers for quaternions. The Hurwitz quaternions consist of all quaternions with integer components and all quaternions with half-integer components. These two sets can be combined into a single formula

$$\alpha = \frac{1}{2}E_0(1+\mathbf{i}+\mathbf{j}+\mathbf{k})+E_1\mathbf{i}+E_2\mathbf{j}+E_3\mathbf{k} = a_0 + a_1\mathbf{i}+a_2\mathbf{j}+a_3\mathbf{k}$$

where E_0, E_1, E_2, E_3 are integers. Thus, the quaternion components a_0, a_1, a_2, a_3 are either all integers or all half-integers, depending on whether E_0 is even or odd, respectively. The set of Hurwitz quaternions forms a ring; in particular, the sum or product of any two Hurwitz quaternions is likewise a Hurwitz quaternion.

The generalized Euclidean algorithm identifies the greatest common right or left divisor of two Hurwitz quaternions, where the "size" of the remainder ρ is measured by the norm. The norm $\mathrm{N}(\alpha)$ of a quaternion α is the nonnegative real number

$$\mathrm{N}(\alpha) = \alpha\bar{\alpha} = a_0^2 + a_1^2 + a_2^2 + a_3^2$$

where $\bar{\alpha} = a_0 - a_1\mathbf{i} - a_2\mathbf{j} - a_3\mathbf{k}$ is the conjugate of α.

Since quaternion multiplication is associative, and real numbers commute with other quaternions, the norm of a product of quaternions equals the product of the norms:

$$\mathrm{N}(\alpha\beta) = \alpha\beta\overline{(\alpha\beta)} = \alpha\beta\bar{\beta}\bar{\alpha} = \alpha\mathrm{N}(\beta)\bar{\alpha} = \alpha\bar{\alpha}\mathrm{N}(\beta) = \mathrm{N}(\alpha)\mathrm{N}(\beta)$$

The ring of Hurwitz quaternions is Euclidean, since for any quaternion $\alpha = a_0 + a_1\mathbf{i} + a_2\mathbf{j} + a_3\mathbf{k}$ with rational coefficients we can choose a Hurwitz quaternion $\beta = b_0 + b_1\mathbf{i} + b_2\mathbf{j} + b_3\mathbf{k}$ so that $\mathrm{N}(\alpha - \beta) < 1$ by first choosing b_0 so that $|a_0 - b_0| \leq 1/4$ and then b_1, b_2, b_3 so that $|a_i - b_i| \leq 1/2$ for *i*1, 2, 3. Then we obtain

$$\mathrm{N}(\alpha - \beta) = (a_0 - b_0)^2 + (a_1 - b_1)^2 + (a_2 - b_2)^2 + (a_3 - b_3)^2$$

$$\leq \left(\frac{1}{4}\right)^2 + \left(\frac{1}{2}\right)^2 + \left(\frac{1}{2}\right)^2 + \left(\frac{1}{2}\right)^2 = \frac{13}{16} < 1.$$

Therefore, the ring of Hurwitz quaternions is Euclidean and, as a consequence, also a unique factorization domain.

Since any natural number can be factored into powers of primes, the four-square theorem therefore holds for all natural numbers if it is true for all prime numbers. It is true for $2 = 0^2 + 0^2 + 1^2 + 1^2$. To show this for an odd prime integer p, represent it as a quaternion $(p, 0, 0, 0)$ and assume that it can be factored into two Hurwitz quaternions

$$p = \alpha\beta.$$

The norms of p, α, β are nonnegative integers such that

$$\mathrm{N}(p) = p^2 = \mathrm{N}(\alpha\beta) = \mathrm{N}(\alpha)\mathrm{N}(\beta).$$

But the norm has only two factors, both p. Therefore, if it can be factored into Hurwitz quaternions, then p is the sum of four squares

$$p = \mathrm{N}(\alpha) = a_0^2 + a_1^2 + a_2^2 + a_3^2.$$

Lagrange proved that any odd prime p divides at least one number of the form $1 + l^2 + m^2$, where l and m are integers. The latter number can be factored in Hurwitz quaternions:

$$1 + l^2 + m^2 = (1 + l\,\mathrm{i} + m\,\mathrm{j})\,(1 - l\,\mathrm{i} - m\,\mathrm{j}).$$

If p could not be factored in Hurwitz quaternions, it would be a Hurwitz prime number by definition. Then, by unique factorization, p would have to divide either $1 + l\,\mathrm{i} + m\,\mathrm{j}$ or $1 - l\,\mathrm{i} - m\,\mathrm{j}$ to form another Hurwitz quaternion. But for $p>2$, the number

$$1/p \pm l/p\,\mathrm{i} \pm m/p\,\mathrm{j}$$

is not a Hurwitz integer. Therefore, every $p>2$ can be factored in Hurwitz quaternions, and the four-square theorem holds.

Generalizations

Lagrange's four-square theorem is a special case of the Fermat polygonal number theorem and Waring's problem. Another possible generalisation is the following problem: Given natural numbers *a, b, c, d,* can we solve

$$n = ax_1^2 + bx_2^2 + cx_3^2 + dx_4^2$$

for all positive integers n in integers x_1, x_2, x_3, x_4? The case $a = b = c = d = 1$ is answered in the positive by Lagrange's four-square theorem. The general solution was given by Ramanujan. He proved that if we assume, without loss of generality, that $a \leq b \leq c \leq d$ then there are exactly 54 possible choices for *a, b, c, d* such that the problem is solvable in integers x_1, x_2, x_3, x_4 for all n. (Ramanujan listed a 55th possibility $a = 1$, $b = 2$, $c = 5$, $d = 5$, but in this case the problem is not solvable if $n = 15$.)

Algorithms

Michael O. Rabin and Jeffrey Shallit have found randomized polynomial-time algorithms for computing a representation $n = x_1^2 + x_2^2 + x_3^2 + x_4^2$ for a given integer n, in expected running time $O((\log n)^2)$.

Uniqueness

The sequence of positive integers whose representation as a sum of four squares is unique (up to order) is:

1, 2, 3, 5, 6, 7, 8, 11, 14, 15, 23, 24, 32, 56, 96, 128, 224, 384, 512, 896 ... (sequence A006431 in OEIS).

These integers consist of the seven odd numbers 1, 3, 5, 7, 11, 15, 23 and all numbers of the form 2×4^k, 6×4^k or 14×4^k.

The sequence of positive integers which cannot be represented as a sum of four *non-zero* squares is:

1, 2, 3, 5, 6, 8, 9, 11, 14, 17, 24, 29, 32, 41, 56, 96, 128, 224, 384, 512, 896 ... (sequence A000534 in OEIS).

These integers consist of the eight odd numbers 1, 3, 5, 9, 11, 17, 29, 41 and all numbers of the form 2×4^n, 6×4^n or 14×4^n.

Lambek–Moser Theorem

In combinatorial number theory, the Lambek–Moser theorem is a generalization of Beatty's theorem that defines a partition of the positive integers into two subsets from any monotonic integer-valued function. Conversely, any partition of the positive integers into two subsets may be defined from a monotonic function in this way.

The theorem was discovered by Leo Moser and Joachim Lambek. Dijkstra (1980) provides a visual proof of the result.Statement of the theorem

The theorem applies to any non-decreasing and unbounded function f that maps positive integers to non-negative integers. From any such function f, define f^* to be the integer-valued function that is as close as possible to the inverse function of f, in the sense that, for all n,

$f(f^*(n)) < n \leq f(f^*(n) + 1)$. It follows from this definition that $f^{**} = f$.

Further, define

$F(n) = f(n) + n$ and $G(n) = f^*(n) + n$.

Then the result states that F and G are strictly increasing and that the ranges of F and G form a partition of the positive integers.

Example

Let $f(n) = n^2$;[2] then $f^*(n) = \lfloor \sqrt{n-1} \rfloor$. Thus $F(n) = n^2 + n$ and $G(n) = \lfloor \sqrt{n-1} \rfloor + n$. For $n = 1, 2, 3, \ldots$ the values of F are the pronic numbers

2, 6, 12, 20, 30, 42, 56, 72, 90, 110, ...

while the values of G are

1, 3, 4, 5, 7, 8, 9, 10, 11, 13, 14, ...

These two sequences are complementary: each positive integer belongs to exactly one of them.

The Lambek–Moser theorem states that this phenomenon is not specific to the pronic numbers, but rather it arises for any choice of f with the appropriate properties.

Universality

The Lambek–Moser theorem is universal, in the sense that it can explain any partition of the integers into two infinite parts. If $S = s_1, s_2, \ldots$ and $T = t_1, t_2, \ldots$ are any two infinite subsets forming a partition of the integers, one may construct a pair of functions f and f^* from which this partition may be derived using the Lambek–Moser theorem: define $f(n) = s_n - n$ and $f^*(n) = t_n - n$.

For instance, consider the partition of integers into even and odd numbers: let S be the even numbers and T be the odd numbers. Then $s_n = 2n$, so $f(n) = n$ and similarly $f^*(n) = n - 1$. These two functions f and f^* form an inverse pair, and the partition generated via the Lambek–Moser theorem from this pair is just the partition of the positive integers into even and odd numbers.

Lambek and Moser discuss formulas involving the prime-counting function for the functions f and f^* arising in this way from the partition of the positive integers into prime numbers and composite numbers.

Landsberg–Schaar Relation

In number theory and harmonic analysis, the Landsberg–Schaar relation (or identity) is the following equation, which is valid for arbitrary positive integers p and q:

$$\frac{1}{\sqrt{p}}\sum_{n=0}^{p-1}\exp\left(\frac{2\pi i n^2 q}{p}\right)=\frac{e^{\pi i/4}}{\sqrt{2q}}\sum_{n=0}^{2q-1}\exp\left(-\frac{\pi i n^2 p}{2q}\right).$$

Although both sides are mere finite sums, no proof by entirely finite methods has yet been found. The standard way to prove it is to put $\tau = 2iq/p+\varepsilon$, where $\varepsilon > 0$ in this identity due to Jacobi (which is essentially just a special case of the Poisson summation formula in classical harmonic analysis):

$$\sum_{n=-\infty}^{+\infty} e^{-\pi n^2 \tau} = \frac{1}{\sqrt{\tau}}\sum_{n=-\infty}^{+\infty} e^{-\pi n^2/\tau}$$

and then let $\varepsilon \to 0$.

If we let $q = 1$, the identity reduces to a formula for the quadratic Gauss sum modulo p. The Landsberg–Schaar identity can be rephrased more symmetrically as

$$\frac{1}{\sqrt{p}}\sum_{n=0}^{p-1}\exp\left(\frac{\pi i n^2 q}{p}\right)=\frac{e^{\pi i/4}}{\sqrt{q}}\sum_{n=0}^{q-1}\exp\left(-\frac{\pi i n^2 p}{q}\right)$$

provided that we add the hypothesis that pq is an even number.

Lindemann–Weierstrass Theorem

In mathematics, the Lindemann–Weierstrass theorem is a result that is very useful in establishing the transcendence of numbers. It states that if $\alpha_1, ..., \alpha_n$ are algebraic numbers which are linearly independent over the rational numbers Q, then $e^{\alpha_1}, ..., e^{\alpha_n}$ are algebraically independent over Q; in other words the extension field $Q(e^{\alpha_1}, ..., e^{\alpha_n})$ has transcendence degree n over Q.

An equivalent formulation (Baker 1975, Chapter 1, Theorem 1.4), is the following: If $\alpha_1, ..., \alpha_n$ are distinct algebraic numbers, then the exponentials $e^{\alpha_1}, ..., e^{\alpha_n}$ are linearly independent over the algebraic numbers. This equivalence transforms a linear relation over the algebraic numbers into an algebraic relation over the rational numbers by using the fact that a symmetric polynomial whose arguments are all conjugates of one another gives a rational number.

The theorem is named for Ferdinand von Lindemann and Karl Weierstrass. Lindemann proved in 1882 that e^{α} is transcendental for every non-zero algebraic number α, thereby establishing that pi (π) is transcendental. Weierstrass proved the above more general statement in 1885.

The theorem, along with the Gelfond–Schneider theorem, is generalized by Schanuel's conjecture.

Naming Convention

The theorem is also known variously as the Hermite–Lindemann theorem and the Hermite–Lindemann–Weierstrass theorem. Charles Hermite first proved the simpler theorem where the α_i exponents are required to be rational integers and linear independence is only assured over the rational integers, a result sometimes referred to as Hermite's theorem. Although apparently a rather special case of the above theorem, the general result can be reduced to this simpler case. Lindemann was the first to allow algebraic numbers into Hermite's work in 1882. Shortly afterwards

Weierstrass obtained the full result, and further simplifications have been made by several mathematicians, most notably by David Hilbert.

Transcendence of e and π

The transcendence of e and π are direct corollaries of this theorem.

Suppose α is a nonzero algebraic number; then α is a linearly independent set over the rationals, and therefore by the first formulation of the theorem $\{e^{\alpha}\}$ is an algebraically independent set; or in other words e^{α} is transcendental. In particular, $e^1 = e$ is transcendental. (A more elementary proof that e is transcendental is outlined in the article on transcendental numbers.)

Alternatively, by the second formulation of the theorem, if α is a nonzero algebraic number, then $\{0, \alpha\}$ is a set of distinct algebraic numbers, and so the set $\{e^0, e^{\alpha}\} = \{1, e^{\alpha}\}$ is linearly independent over the algebraic numbers and in particular e^{α} cannot be algebraic and so it is transcendental.

The proof that π is transcendental is by contradiction. If π were algebraic, πi would be algebraic as well, and then by the Lindemann–Weierstrass theorem $e^{\pi i}$ = "1 would be transcendental, a contradiction.

A slight variant on the same proof will show that if α is a nonzero algebraic number then $\sin(\alpha)$, $\cos(\alpha)$, $\tan(\alpha)$ and their hyperbolic counterparts are also transcendental.

p-adic Conjecture

The p-adic Lindemann–Weierstrass conjecture is that a p-adic analog of this statement is also true: Suppose p is some prime number and $\alpha_1, \ldots, \alpha_n$ are p-adic numbers which are algebraic over Q and linearly independent over Q, such that $|\alpha_i|_p < 1/p$ for all i; then the p-adic exponentials $\exp_p(\alpha_1), \ldots, \exp_p(\alpha_n)$ are p-adic numbers that are algebraically independent over Q.

Modular Conjecture

An analogue of the theorem involving the modular function j was conjectured by Daniel Bertrand in 1997, and remains an open problem. Writing $q = e^{2\pi i\tau}$ for the nome and $j(\tau) = J(q)$, the conjecture is as follows. Let $q_1,\ldots,q_n$ be non-zero algebraic numbers in the complex unit disc such that the $3n$ numbers $J(q_i)$, $J^2(q_i)$, $J^{2\,2}(q_i)$ $(1 \le i \le n)$ are algebraically dependent over the rational numbers. Then there exist two indices $1 \le i < j \le n$ such that q_i and q_j are multiplicatively dependent.

Proof of the Lindemann–Weierstrass Theorem

We prove Baker's reformulation of the theorem: that if a_1, a_2... a_n are non-zero algebraic numbers, and α_1, α_2... α_n are distinct algebraic numbers, then $a_1e^{\alpha_1} + a_2e^{\alpha_2} + \cdots + a_ne^{\alpha_n} \neq 0$.

6

Linear Congruence Theorem

In modular arithmetic, the question of when a linear congruence can be solved is answered by the linear congruence theorem. If a and b are any integers and n is a positive integer, then the congruence

$$ax \equiv b \pmod{n}$$

has a solution for x if and only if b is divisible by the greatest common divisor d of a and n (denoted by gcd(a,n)). When this is the case, and x_0 is one solution of (1), then the set of all solutions is given by

$$\{x_0 + k\frac{n}{d} \mid k \in \mathbb{Z}\}.$$

In particular, there will be exactly d = gcd(a,n) solutions in the set of residues {0,1,2,...,n-1}.

For example, examining the equation ax a– 2 (mod 6) with different values of a yields

$$3x \equiv 2 \pmod{6}$$

Here d = gcd(3,6) = 3 but since 3 does not divide 2, there is no solution.

$$5x \equiv 2 \pmod{6}$$

Here d = gcd(5,6) = 1, which divides any b, and so there is just one solution in {0,1,2,3,4,5}: x=4.

$$4x \equiv 2 \pmod{6}$$

Here d = gcd(4,6) = 2, which does divide 2, and so there are exactly two solutions in {0,1,2,3,4,5}: x=2 and x=5.

Solving a Linear Congruence

In general solving equations of the form:

$$ax \equiv b \pmod{n}$$

If the greatest common divisor d = gcd(a, n) divides b, then we can find a solution x to the congruence as follows: the extended Euclidean algorithm yields integers r and s such $ra + sn = d$. Then $x = rb/d$ is a solution. The other solutions are the numbers congruent to x modulo n/d.

For example, the congruence

$$12x \equiv 20 \pmod{28}$$

has 4 solutions since gcd(12, 28) = 4 divides 20. The extended Euclidean algorithm gives (-2)*12 + 1*28 = 4, i.e. r = -2 and s = 1. Therefore, one solution is x = -2*20/4 = -10, and -10 = 4 modulo 7. All other solutions will also be congruent to 4 modulo 7. Since the original equation uses modulo 28, the entire solution set in the range from 0 to 27 is x = {4,11,18,25}

System of Linear Congruences

By repeatedly using the linear congruence theorem, one can also solve systems of linear congruences, as in the following example: find all numbers x such that

2x a– 2 (mod 6)

3x a– 2 (mod 7)

2x a– 4 (mod 8).

By solving the first congruence using the method explained above, we find x a– 1 (mod 3), which can also be written as $x = 3k + 1$. Substituting this into the second congruence and simplifying, we get

9k a□ □“1 (mod 7).

Solving this congruence yields k a– 3 (mod 7), or $k = 7l + 3$. It then follows that $x = 3\ (7l + 3) + 1 = 21l + 10$. Substituting this into the third congruence and simplifying, we get

42l a– “16 (mod 8)

which has the solution l a– 0 (mod 4), or $l = 4m$. This yields x =

21(4*m*) + 10 = 84*m* + 10, or

$$x \text{ a}- 10 \pmod{84}$$

which describes all solutions to the system.

Lochs' Theorem

In number theory, Lochs' theorem is a theorem concerning the rate of convergence of the continued fraction expansion of a typical real number. The theorem was proved by Gustav Lochs in 1964.

The theorem states that for almost all real numbers in the interval (0,1), the number of terms *m* of the number's continued fraction expansion that are required to determine the first *n* places of the number's decimal expansion behaves asymptotically as follows:

$$\lim_{n\to\infty}\frac{m}{n}=\frac{6\ln 2\ln 10}{\pi^2}\approx 0.97027014 \text{ (sequence A086819 in OEIS).}$$

As this limit is only slightly smaller than 1, this can be interpreted as saying that each additional term in the continued fraction representation of a "typical real number increases the accuracy of the representation by approximately one decimal place. The decimal system is the last positional system for which each digit carries less information than one continued fraction quotient; going to base-11 (changing to in the equation) makes the above value exceed 1.

The reciprocal of this limit,

$$\frac{\pi^2}{6\ln 2\ln 10}\approx 1.03064083 \text{ (sequence A062542 in OEIS),}$$

is twice the base-10 logarithm of Lévy's constant.

Mahler's Compactness Theorem

In mathematics, Mahler's compactness theorem, proved by Kurt Mahler (1946), is a foundational result on lattices in Euclidean space, characterising sets of lattices that are 'bounded' in a certain definite sense. Looked at another way, it explains the ways in which a lattice could degenerate (*go off to infinity*) in a sequence of lattices. In intuitive terms it says that this is possible in just two ways: becoming *coarse-grained* with a fundamental domain that has ever larger volume; or containing shorter and shorter vectors. It is also called his selection theorem, following an older convention used in naming compactness theorems, because they were formulated in terms of sequential compactness (the possibility of selecting a convergent subsequence).

Let X be the space

$$GL_n(R)/GL_n(Z)$$

that parametrises lattices in R^n, with its quotient topology. There is a well-defined function Δ on X, which is the absolute value of the determinant of a matrix — this is constant on the cosets, since an invertible integer matrix has determinant 1 or "1.

Mahler's compactness theorem states that a subset Y of X is relatively compact if and only if Δ is bounded on Y, and there is a neighbourhood N of $\{0\}$ in R^n such that for all $\wedge$ in Y, the only lattice point of $\wedge$ in N is $\{0\}$ itself.

The assertion of Mahler's theorem is equivalent to the compactness of the space of unit-covolume lattices in whose systole is larger than .

Mahler's compactness theorem was generalized to semisimple Lie groups by Mumford.

Schnirelmann Density

In additive number theory, the Schnirelmann density of a sequence of numbers is a way to measure how "dense" the sequence is. It is named after Russian mathematician L.G. Schnirelmann, who was the first to study it.

Definition

The Schnirelmann density of a set of natural numbers A is defined as

$$\sigma A = \inf_n \frac{A(n)}{n},$$

where $A(n)$ denotes the number of elements of A not exceeding n and inf is infimum.

The Schnirelmann density is well-defined even if the limit of $A(n)/n$ as $n \to$ " fails to exist.

Properties

By definition, $0 \leq A(n) \leq$ n and $n\ \sigma A \leq A(n)$ for all n, and therefore $0 \leq \sigma A \leq 1$, and $\sigma A = 1$ if and only if $A = \mathrm{N}$. Furthermore,

$$\sigma A = 0 \Rightarrow \forall \epsilon > 0\ \exists n\ A(n) < \epsilon n.$$

Sensitivity

The Schnirelmann density is sensitive to the first values of a set:

$$\forall k\ k \notin A \Rightarrow \sigma A \leq 1 - 1/k.$$

In particular,

$$1 \notin A \Rightarrow \sigma A = 0$$

and

$$2 \notin A \Rightarrow \sigma A \leq \frac{1}{2}.$$

Consequently, the Schnirelmann densities of the even numbers and the odd numbers, which one might expect to agree, are 0 and 1/2 respectively. Schnirelmann and Yuri Linnik exploited this sensitivity as we shall see.

Schnirelmann's Theorems

If we set $\mathfrak{G}^2 = \{k^2\}_{k=1}^{\infty}$, then Lagrange's four-square theorem can be restated as $\sigma(\mathfrak{G}^2 \oplus \mathfrak{G}^2 \oplus \mathfrak{G}^2 \oplus \mathfrak{G}^2) = 1$. (Here the $A \oplus B$ symbol denotes the sumset of $A \cup \{0\}$ and $B \cup \{0\}$.) It is clear that $\sigma \mathfrak{G}^2 = 0$. In fact, we still have $\sigma(\mathfrak{G}^2 \oplus \mathfrak{G}^2) = 0$, and one might ask at what point the sumset attains Schnirelmann density 1 and how does it increase. It actually is the case that $\sigma(\mathfrak{G}^2 \oplus \mathfrak{G}^2 \oplus \mathfrak{G}^2) = 5/6$ and $\mathfrak{G}^2$ one sees that sumsetting once again yields a more populous set, namely all of $\mathbb{N}$. Schnirelmann further succeeded in developing these ideas into the following theorems, aiming towards Additive Number Theory, and proving them to be a novel resource (if not greatly powerful) to attack important problems, such as Waring's problem and Goldbach's conjecture.

Theorem. Let A and B be subsets of $\mathbb{N}$. Then

$$\sigma(A \oplus B) \geq \sigma A + \sigma B - \sigma A \cdot \sigma B$$

Note that $\sigma A + \sigma B - \sigma A \cdot \sigma B = 1 - (1 - \sigma A)(1 - \sigma B)$. Inductively, we have the following generalization.

Corollary. Let $A_i \subseteq \mathbb{N}$ be a finite family of subsets of $\mathbb{N}$. Then

$$\sigma(\bigoplus_i A_i) \geq 1 - \prod_i (1 - \sigma A_i)$$

The theorem provides the first insights on how sumsets accumulate. It seems unfortunate that its conclusion stops short of showing σ being superadditive. Yet, Schnirelmann provided us with the following results, which sufficed for most of his purpose.

Theorem. Let A and B be subsets of $\mathbb{N}$. If $\sigma A + \sigma B \geq 1$, then

$$A \oplus B = \mathbb{N}$$

Theorem. (*Schnirelmann*) Let $A \subseteq \mathbb{N}$. If $\sigma A > 0$ then there exists such that

$$\bigoplus_{i=1}^{k} A = \mathbb{N}$$

Additive Bases

A subset $A \subseteq \mathbb{N}$ with the property that $A \oplus A \oplus \cdots \oplus A = \mathbb{N}$ for a finite sum, is called an additive basis, and the least number of summands required is called the *degree* (sometimes *order*) of the basis. Thus, the last theorem states that any set with positive Schnirelmann density $\mathfrak{G}^2 = \{k^2\}_{k=1}^{\infty}$ is an additive basis. In this terminology, the set of squares is an additive basis of degree 4.

Schnirelmann's Theorem

In 1931 Schnirelmann used these ideas in conjunction with the Brun sieve to prove Schnirelmann's theorem, that any natural number greater than one can be written as the sum of not more than C prime numbers, where C is an effectively computable constant. Schnirelmann's constant is the lowest number C with this property.

Olivier Ramaré showed in (Ramaré 1995) that Schnirelmann's constant is at most 7, improving the earlier upper bound 19 by Hans Riesel and R. C. Vaughan.

Schnirelmann's constant is at least 3; Goldbach's conjecture implies that this is the constant's actual value.

Essential Components

Khintchin proved that the sequence of squares, though of zero Schnirelmann density, when added to a sequence of Schnirelmann density between 0 and 1, increases the density:

$$\sigma(A+\mathfrak{G}^2)>\sigma(A) \text{ for } 0<\sigma(A)<1.$$

This was soon simplified and extended by Erdõs, who showed, that if A is any sequence with Schnirelmann density α and B is an additive basis of order k then

$$\sigma(A+B)\geq\alpha+\frac{\alpha(1-\alpha)}{2k}.$$

Sequences with this property were named essential components by Khintchin. Linnik showed that an essential component need not be an additive basis as he constructed an essential component that has $x^{o(1)}$ elements less than x. More precisely, the sequence has

$$e^{(\log x)^c}$$

elements less than x for some $c<1$. This was improved by E. Wirsing to

$$e^{\sqrt{\log x}\log\log x}.$$

For a while, it remained an open problem how many elements an essential component must have. Finally, Ruzsa determined that an essential component has at least $(\log x)^c$ elements up to x, for some $c>1$, and for every $c>1$ there is an essential component which has at most $(\log x)^c$ elements up to x.

Diophantine Set

In mathematics, a Diophantine equation is an equation of the form $P(x_1, ..., x_j, y_1, ..., y_k)=0$ (usually abbreviated $P(x,y)=0$) where $P(x,y)$ is a polynomial with integer coefficients. A Diophantine set is a subset S of N^j so that for some Diophantine equation $P(x,y)=0$.

$$\bar{n}\in S \Leftrightarrow (\exists\bar{m}\in\mathbb{N}^k)(P(\bar{n},\bar{m})=0)$$

That is, a parametre value is in the Diophantine set S if and only if the associated Diophantine equation is satisfiable under that parametre value. Note that the use of natural numbers both in S and the existential quantification merely reflects the usual applications in computability and model theory. We can equally well speak of Diophantine sets of integers and freely replace quantification over natural numbers with quantification over the integers.

Also it is sufficient to assume P is a polynomial over and multiply P by the appropriate denominators to yield integer coefficients. However, whether quantification over rationals can also be substituted for quantification over the integers it is a notoriously hard open problem.

The MRDP theorem states that a set of integers is Diophantine if and only if it is computably enumerable. A set S is recursively enumerable precisely if there is an algorithm that, when given an integer, eventually halts if that input is a member of S and otherwise runs forever. This means that the concept of general Diophantine set, apparently belonging to number theory, can be taken rather in logical or recursion-theoretic terms. This is far from obvious, however, and represented the culmination of some decades of work.

Matiyasevich's completion of the MRDP theorem settled Hilbert's tenth problem. Hilbert's tenth problem was to find a general algorithm which can decide whether a given Diophantine equation has a solution among the integers. While Hilbert's tenth problem is not a formal mathematical statement as such the nearly universal acceptance of the (philosophical) identification of a decision algorithm with a total computable predicate allows us to use the MRDP theorem to conclude the tenth problem is unsolvable.

Examples

The well known Pell equation

$$x^2 - d(y+1)^2 = 1$$

is an example of a Diophantine equation with a parametre. As has long been known, the equation has a solution in the unknowns x, y precisely when the parametre is d 0 or not a perfect square. In the present context, one says that this equation provides a *Diophantine definition* of the set

$$\{0,2,3,5,6,7,8,10,...\}$$

consisting of 0 and the natural numbers that are not perfect squares. Other examples of Diophantine definitions are as follows:

- The equation $a = (2x+3)y$ only has solutions $\mathbb{N}$ in when a is not a power of 2.
- The equation $a = (x+2)(y+2)$ only has solutions in $\mathbb{N}$ when a is greater than 1 and is not a prime number.
- The $a + x = b$ equation defines the set of pairs (a,b) such that $a \le b$.

Matiyasevich's Theorem

Matiyasevich's theorem says: Every computably enumerable set is Diophantine. A set S of integers is computably enumerable if there is an algorithm that behaves as follows: When given as input an integer n, if n is a member of S, then the algorithm eventually halts; otherwise it runs forever. That is equivalent to saying there is an algorithm that runs forever and lists the members of S. A set S is Diophantine precisely if there is some polynomial with integer coefficients $f(n, x_1, ..., x_k)$ such that an integer n is in S if and only if there exist some integers $x_1, ..., x_k$ such that $f(n, x_1, ..., x_k) = 0$.

It is not hard to see that every Diophantine set is recursively enumerable: consider a Diophantine equation $f(n, x_1, ..., x_k) = 0$. Now we make an algorithm which simply tries all possible values for $n, x_1, ..., x_k$, in the increasing order of the sum of their absolute values, and prints n every time $f(n, x_1, ..., x_k) = 0$. This algorithm will obviously run forever and will list exactly the n for which $f(n, x_1, ..., x_k) = 0$ has a solution in $x_1, ..., x_k$.

Proof Technique

Yuri Matiyasevich utilized a method involving Fibonacci numbers in order to show that solutions to Diophantine equations may grow exponentially. Earlier work by Julia Robinson, Martin Davis and Hilary Putnam had shown that this suffices to show that every computably enumerable set is Diophantine.

Application to Hilbert's Tenth Problem

Hilbert's tenth problem asks for a general algorithm deciding the solvability of Diophantine equations. The conjunction of

Matiyasevich's theorem with earlier results known collectively as the MRDP theorem implies that a solution to Hilbert's tenth problem is impossible.

Refinements

Later work has shown that the question of solvability of a Diophantine equation is undecidable even if the equation only has 9 natural number variables (Matiyasevich, 1977) or 11 integer variables (Zhi Wei Sun, 1992).

Further Applications

Matiyasevich's theorem has since been used to prove that many problems from calculus and differential equations are unsolvable.

One can also derive the following stronger form of Gödel's first incompleteness theorem from Matiyasevich's result:

Corresponding to any given consistent axiomatization of number theory, one can explicitly construct a Diophantine equation which has no solutions, but such that this fact cannot be proved within the given axiomatization.

Mazur's Torsion Theorem

In algebraic geometry Mazur's torsion theorem, due to Barry Mazur, classifies the possible torsion subgroups of the group of rational points on an elliptic curve defined over the rational numbers.

If C_n denotes the cyclic group of order n, then the possible torsion subgroups are C_n with $1 \leq n \leq 10$, and also C_{12}; and the direct sum of C_2 with C_2, C_4, C_6 or C_8.

In the opposite direction, all these torsion structures occur infinitely often over Q since the corresponding modular curves are of genus zero.

Meyer's Theorem

In number theory, Meyer's theorem on quadratic forms states that an indefinite quadratic form Q in five or more variables over the field of rational numbers nontrivially represents zero. In other words, if the equation

$$Q(x) = 0$$

has a non-zero real solution, then it has a non-zero rational solution (the converse is obvious). By clearing the denominators, an integral solution x may also be found.

Meyer's theorem is usually from the Hasse–Minkowski theorem (which was later proved) and the following statement:

A rational quadratic form in five or more variables represents zero over the field Q_p of the p-adic numbers for all p.

Meyer's theorem is best possible with respect to the number of variables: there are indefinite rational quadratic forms Q in four variables which do not represent zero. One family of examples is given by

$Q(x_1,x_2,x_3,x_4) = x_1^2 + x_2^2$ " $p(x_3^2 + x_4^2)$,

where p is a prime number that is congruent to 3 modulo 4. This can be proved by the method of infinite descent using the fact that if the sum of two perfect squares is divisible by such a p then each summand is divisible by p.

Midy's Theorem

In mathematics, Midy's theorem, named after French mathematician E. Midy, is a statement about the decimal expansion of fractions a/p where p is a prime and a/p has a repeating decimal expansion with an even period. If the period of the decimal representation of a/p is $2n$, so that

$$\frac{a}{p} = 0.\overline{a_1 a_2 a_3 \ldots a_n a_{n+1} \ldots a_{2n}}$$

then the digits in the second half of the repeating decimal period are the 9s complement of the corresponding digits in its first half. In other words

$$a_i + a_{i+n} = 9$$

$$a_1 \ldots a_n + a_{n+1} \ldots a_{2n} = 10^n - 1.$$

For example

$$\frac{1}{17} = 0.\overline{0588235294117647} \text{ and } 05882352$$

$$+94117647 = 99999999.$$

Midy's Theorem in Other Bases

Midy's theorem and its extension do not depend on special properties of the decimal expansion, but work equally well in any base b, provided we replace $10^k - 1$ with $b^k - 1$ and carry out addition in base b. For example, in octal

$$\frac{1}{19} = 0.\overline{032745}_8$$

$$032_8 + 745_8 = 777_8$$

$$03_8 + 27_8 + 45_8 = 77_8.$$

Proof of Midy's Theorem

Short proofs of Midy's theorem can be given using results from group theory. However, it is also possible to prove Midy's theorem using elementary algebra and modular arithmetic:

Let p be a prime and a/p be a fraction between 0 and 1. Suppose the expansion of a/p in base b has a period of –!, so

$$\frac{a}{p} = [0.\overline{a_1 a_2 \ldots a_\ell}]_b$$

$$[6pt] \quad \Rightarrow \frac{a}{p} b^\ell = [a_1 a_2 \ldots a_\ell . \overline{a_1 a_2 \ldots a_\ell}]_b$$

$$[6pt] \quad \Rightarrow \frac{a}{p} b^\ell = N + [0.\overline{a_1 a_2 \ldots a_\ell}]_b = N + \frac{a}{p}$$

$$[6pt] \quad \Rightarrow \frac{a}{p} = \frac{N}{b^\ell - 1}$$

where N is the integer whose expansion in base b is the string $a_1 a_2 \ldots a_{-!}$.

Note that $b^{-!} - 1$ is a multiple of p because $(b^{-!} - 1)a/p$ is an integer. Also $b^n - 1$ is *not* a multiple of p for any value of n less than –!, because otherwise the repeating period of a/p in base b would be less than –!.

Now suppose that $-! = hk$. Then $b^{-!} - 1$ is a multiple of $b^k - 1$. Say $b^{-!} - 1 = m(b^k - 1)$, so

$$\frac{a}{p} = \frac{N}{m(b^k - 1)}.$$

But $b^{-!} - 1$ is a multiple of p; $b^k - 1$ is *not* a multiple of p (because k is less than –!); and p is a prime; so m must be a multiple of p and

$$\frac{am}{p} = \frac{N}{b^k - 1}$$

is an integer. In other words

$$N \equiv 0 \pmod{b^k - 1}.$$

Now split the string $a_1a_2...a_{-!}$ into h equal parts of length k, and let these represent the integers $N_0...N_{h-1}$ in base b, so that

$$\begin{aligned} N_{h-1} &= [a_1 \dots a_k]_b \\ N_{h-2} &= [a_{k+1} \dots a_{2k}]_b \\ &\vdots \\ N_0 &= [a_{l-k+1} \dots a_l]_b \end{aligned}$$

To prove Midy's extended theorem in base b we must show that the sum of the h integers N_i is a multiple of $b^k - 1$.

Since b^k is congruent to 1 modulo $b^k - 1$, any power of b^k will also be congruent to 1 modulo $b^k - 1$. So

$$N = \sum_{i=0}^{h-1} N_i b^{ik} = \sum_{i=0}^{h-1} N_i (b^k)^i$$

$$\Rightarrow N \equiv \sum_{i=0}^{h-1} N_i \pmod{b^k - 1}$$

$$\Rightarrow \sum_{i=0}^{h-1} N_i \equiv 0 \pmod{b^k - 1}$$

which proves Midy's extended theorem in base b.

To prove the original Midy's theorem, take the special case where $h = 2$. Note that N_0 and N_1 are both represented by strings of k digits in base b so both satisfy

$$0 \le N_i \le b^k - 1.$$

N_0 and N_1 cannot both equal 0 (otherwise $a/p = 0$) and cannot both equal $b^k - 1$ (otherwise $a/p = 1$), so

$$0 < N_0 + N_1 < 2(b^k - 1)$$

and since $N_0 + N_1$ is a multiple of $b^k - 1$, it follows that

$$N_0 + N_1 = b^k - 1.$$

Catalan's Conjecture

Catalan's conjecture (or Mihãilescu's theorem) is a theorem in number theory that was conjectured by the mathematician Eugène Charles Catalan in 1844 and proven in 2002 by Preda Mihãilescu.

2^3 and 3^2 are two powers of natural numbers, whose values 8 and 9 respectively are consecutive. The conjecture states that this is the *only* case of two consecutive powers. That is to say, that the only solution in the natural numbers of

$$x^a \text{ `` } y^b = 1$$

for $x, a, y, b > 1$ is $x = 3, a = 2, y = 2, b = 3$.

History

The history of the problem dates back at least to Gersonides, who proved a special case of the conjecture in 1343 where x and y were restricted to be 2 or 3.

In 1976, Robert Tijdeman applied methods from the theory of transcendental numbers to show that there is an effectively computable constant C so that the exponents of all consecutive powers are less than C. As the results of a number of other mathematicians collectively had established a bound for the base dependent only on the exponents, this resolved Catalan's conjecture for all but a finite number of cases. However, the finite calculation required to complete the proof of the theorem was nonetheless too time-consuming to perform.

Catalan's conjecture was proved by Preda Mihãilescu in April 2002, so it is now sometimes called *Mihãilescu's theorem*. The proof was published in the *Journal für die reine und angewandte Mathematik*, 2004. It makes extensive use of the theory of cyclotomic fields and Galois modules. An exposition of the proof was given by Yuri Bilu in the Séminaire Bourbaki.

Pillai's conjecture

Pillai's conjecture concerns a general difference of perfect powers (sequence A001597 in OEIS). It states that each positive integer occurs only finitely many times as a difference of perfect powers. It is an open problem and is named for S. S. Pillai.

Paul Erdõs conjectured that there is some positive constant c such that if d is the difference of a perfect power n, then $d>n^c$ for sufficiently large n.

Minkowski's Theorem

In mathematics, Minkowski's theorem is the statement that any convex set in R^n which is symmetric with respect to the origin and with volume greater than 2^n d(L) contains a non-zero lattice point. The theorem was proved by Hermann Minkowski in 1889 and became the foundation of the branch of number theory called the geometry of numbers.

Formulation

Suppose that L is a lattice of determinant d(L) in the n-dimensional real vector space R^n and S is a convex subset of R^n that is symmetric with respect to the origin, meaning that if x is in S then "x is also in S. Minkowski's theorem states that if the volume of S is strictly greater than 2^n d(L), then S must contain at least one lattice point other than the origin.

Example

The simplest example of a lattice is the set Z^n of all points with integer coefficients; its determinant is 1. For $n = 2$ the theorem claims that a convex figure in the plane symmetric about the origin and with area greater than 4 encloses at least one lattice point in addition to the origin. The area bound is sharp: if S is the interior of the square with vertices (±1, ±1) then S is symmetric and convex, has area 4, but the only lattice point it contains is the origin. This observation generalizes to every dimension n.

Proof

The following argument proves Minkowski's theorem for the special case of $L=Z^2$. It can be generalized to arbitrary lattices in arbitrary dimensions.

Consider the map $f: S \to \mathbb{R}^2, (x,y) \mapsto (x \bmod 2, y \bmod 2)$. Intuitively, this map cuts the plane into 2 by 2 squares, then stacks the squares on top of each other. Clearly $f(S)$ has area ≤ 4. Suppose f were injective, which means the pieces of S cut out by the squares stack up in a non-overlapping way. Since f is locally area-preserving, this non-overlapping property would make it area-preserving for all of S, so the area of $f(S)$ would be the same as that of S, which is greater than 4. That is not the case, so f is not injective, and $f(p_1) = f(p_2)$ for some pair of points p_1, p_2 in S. Moreover, we know from the definition of f that $p_2 = p_1 + (2i, 2j)$ for some integers i and j, where i and j are not both zero.

Then since S is symmetric about the origin, $-p_1$ is also a point in S. Since S is convex, the line segment between and lies entirely in S, and in particular the midpoint of that segment lies in S. In other words,

$$\frac{1}{2}\left(-p_1 + p_2\right) = \frac{1}{2}\left(-p_1 + p_1 + (2i, 2j)\right) = (i, j)$$

lies in S. (i,j) is a lattice point, and is not the origin since i and j are not both zero, and so we have found the point we're looking for.

Applications

A corollary of this theorem is the fact that every class in the ideal class group of a number field K contains an integral ideal of norm not exceeding a certain bound, depending on K, called Minkowski's bound.

Modularity theorem

In mathematics the modularity theorem (formerly called the Taniyama–Shimura–Weil conjecture and several related names) states that elliptic curves over the field of rational numbers are related to modular forms. Andrew Wiles proved the modularity theorem for semistable elliptic curves, which was enough to imply Fermat's last theorem, and Christophe Breuil, Brian Conrad, Fred Diamond, and Richard Taylor extended his techniques to prove the full modularity theorem in 2001. The modularity theorem is a special case of more general conjectures due to Robert Langlands. The Langlands programme seeks to attach an automorphic form or automorphic representation (a suitable generalization of a modular form) to more general objects of arithmetic algebraic geometry, such as to every elliptic curve over a number field. Most cases of these extended conjectures have not yet been proved.

Statement

The theorem states that any elliptic curve over Q can be obtained via a rational map with integer coefficients from the classical modular curve

$$X_0(N)$$

for some integer *N*; this is a curve with integer coefficients with an explicit definition. This mapping is called a modular parametrization of level *N*. If *N* is the smallest integer for which such a parametrization can be found (which by the modularity theorem itself is now known to be a number called the *conductor*), then the parametrization may be defined in terms of a mapping generated by a particular kind of modular form of weight two and level *N*, a normalized newform with integer q-expansion, followed if need be by an isogeny.

The modularity theorem implies a closely related analytic statement: to an elliptic curve *E* over Q we may attach a corresponding L-series. The *L*-series is a Dirichlet series, commonly written

$$L(s,E)=\sum_{n=1}^{\infty}\frac{a_n}{n^s}.$$

The generating function of the coefficients is then

$$f(q,E)=\sum_{n=1}^{\infty}a_nq^n.$$

If we make the substitution

$$q=e^{2\pi i\tau}$$

we see that we have written the Fourier expansion of a function $f(\tau,E)$ of the complex variable τ, so the coefficients of the q-series are also thought of as the Fourier coefficients of . The function obtained in this way is, remarkably, a cusp form of weight two and level *N* and is also an eigenform (an eigenvector of all Hecke operators); this is the Hasse–Weil conjecture, which follows from the modularity theorem. Some modular forms of weight two, in turn, correspond to holomorphic differentials for an elliptic curve. The Jacobian of the modular curve can (up to isogeny) be written as a product of irreducible Abelian varieties, corresponding to Hecke eigenforms of weight 2.

The 1-dimensional factors are elliptic curves (there can also be higher dimensional factors, so not all Hecke eigenforms correspond to rational elliptic curves). The curve obtained by finding the corresponding cusp form, and then constructing a curve from it, is isogenous to the original curve (but not, in general, isomorphic to it).

History

Taniyama (1956) stated a preliminary (slightly incorrect) version of the conjecture at the 1955 international symposium on algebraic number theory in Tokyo and Nikko. Goro Shimura and Taniyama worked on improving its rigor until 1957. Weil (1967) rediscovered the conjecture, and showed that it would follow from the (conjectured) functional equations for some twisted L-series of the elliptic curve; this was the first serious evidence that the conjecture might be true. Weil also showed that the conductor of the elliptic curve should be the level of the corresponding modular form.

The conjecture attracted considerable interest when Frey (1986) suggested that the Taniyama–Shimura–Weil conjecture implies Fermat's Last Theorem. He did this by attempting to show that any counterexample to Fermat's Last Theorem would give rise to a non-modular elliptic curve.

However, his argument was not complete. The extra condition which was needed to link Taniyama-Shimura-Weil to Fermat's Last Theorem was identified by Serre (1987) and became known as the epsilon conjecture. In the summer of 1986, Ribet (1990) proved the epsilon conjecture, thereby proving that the Taniyama–Shimura–Weil conjecture implied Fermat's Last Theorem. Wiles (1995), with some help from Richard Taylor, proved the Taniyama–Shimura–Weil conjecture for all semistable elliptic curves, which was strong enough to yield a proof of Fermat's Last Theorem.

The full Taniyama–Shimura–Weil conjecture was finally proved by Diamond (1996), Conrad, Diamond & Taylor (1999), and Breuil et al. (2001) who, building on Wiles' work, incrementally chipped away at the remaining cases until the full result was proved. The now fully proved conjecture became known as the modularity theorem.

Several theorems in number theory similar to Fermat's Last Theorem follow from the modularity theorem. For example: no cube can be written as a sum of two coprime n-th powers, n e– 3. (The case $n = 3$ was already known by Euler.)

Multiplicity-one Theorem

In the mathematical theory of automorphic representations, a multiplicity-one theorem is a result about the representation theory of an adelic reductive algebraic group. The multiplicity in question is the number of times a given abstract group representation is realised in a certain space, of square integrable functions, given in a concrete way.

Definition

Let G be a reductive algebraic group over a number field K and let A denote the adeles of K. Let Z denote the centre of G and let ω be a continuous unitary character from $Z(K)\backslash Z(A)^{\times}$ to $C^{\times}$. Let $L^2{}_0(G(K)/G(A), \omega)$ denote the space of cusp forms with central character ω on G(A). This space decomposes into a direct sum of Hilbert spaces

$$L_0^2(G(K)\setminus G(\mathbf{A}),\omega)=\widehat{\bigoplus}_{(\pi,V_\pi)}m_\pi V_\pi$$

where the sum is over irreducible subrepresentations and m_π are non-negative integers.

The group of adelic points of G, G(A), is said to satisfy the multiplicity-one property if any smooth irreducible admissible representation of G(A) occurs with multiplicity at most one in the space of cusp forms of central character ω, i.e. m_π is 0 or 1 for all such π.

Results

The fact that the general linear group, $GL(n)$, has the multiplicity-one property was proved by Jacquet & Langlands

(1970) for $n = 2$ and independently by Piatetski-Shapiro (1979) and Shalika (1974) for $n > 2$ using the uniqueness of the Whittaker model. Multiplicity-one also holds for $SL(2)$, but not for $SL(n)$ for $n > 2$ (Blasius 1994).

Strong multiplicity One Theorem

The strong multiplicity one theorem of Piatetski-Shapiro (1979) and Jacquet & Shalika (1981) states that two cuspidal automorphic representations of the general linear group are isomorphic if their local components are isomorphic for all but a finite number of places.

Nagell–Lutz Theorem

In mathematics, the Nagell–Lutz theorem is a result in the diophantine geometry of elliptic curves, which describes rational torsion points on elliptic curves over the integers.

Definition of the Terms

Suppose that the equation

$$y^2 = x^3 + ax^2 + bx + c$$

defines a non-singular cubic curve with integer coefficients a, b, c, and let D be the discriminant of the cubic polynomial on the right side:

$$D = -4a^3c + a^2b^2 + 18abc - 4b^3 - 27c^2.$$

Statement of the Theorem

If $P = (x,y)$ is a rational point of finite order on C, for the elliptic curve group law, then:

- 1) x and y are integers
- 2) either $y = 0$, in which case P has order two, or else y divides D, which immediately implies that y^2 divides D.

Generalizations

The Nagell–Lutz theorem generalizes to arbitrary number fields and more general cubic equations.[1]

For curves over the rationals, the generalization says that, for a nonsingular cubic curve whose Weierstrass form

$$y^2 + a_1xy + a_3y = x^3 + a_2x^2 + a_4x + a_6$$

has integer coefficients, any rational point $P=(x,y)$ of finite order must have integer coordinates, or else have order 2 and coordinates of the form $x=m/4$, $y=n/8$, for m and n integers.

History

The result is named for its two independent discoverers, the Norwegian Trygve Nagell (1895–1988) who published it in 1935, and Élisabeth Lutz (1937).

7

Formulas of Numbers

Ostrowski's Theorem

Ostrowski's theorem, due to Alexander Ostrowski (1916), states that any non-trivial absolute value on the rational numbers Q is equivalent to either the usual real absolute value or a *p*-adic absolute value.

Definitions

Two absolute values $|\cdot|$ and $|\cdot|$ on a field $\mathbb{K}$ are defined to be equivalent if there exists a real number $c > 0$ such that

$$|x|_* = |x|^c \quad \text{for all } x \in \mathbb{K}.$$

Observe that this is *stronger* than saying that the two absolute-value structures are topologically isomorphic.

The trivial absolute value on any field $\mathbb{K}$ is defined to be

$$|x|_0 := \begin{cases} 0, & \text{if } x = 0 \\ 1, & \text{if } x \neq 0. \end{cases}$$

The real absolute value on the rationals is the normal absolute value on the reals, defined to be

$$|x|_\infty := \begin{cases} x, & \text{if } x \geq 0 \\ -x, & \text{if } x < 0. \end{cases}$$

This is sometimes written with a subscript 1 instead of infinity.

For a prime number *p*, the *p*-adic absolute value $\mathbb{Q}$ on is defined as follows: any non-zero rational *x*, can be written uniquely as $x = p^n \frac{a}{b}$ with a, b and p pairwise coprime and $n \in \mathbb{Z}$ some integer; so we define

$$|x|_p := \begin{cases} 0, & \text{if } x = 0 \\ p^{-n}, & \text{if } x \neq 0. \end{cases}$$

Another Ostrowski's Theorem

Another theorem states that any field, complete with respect to an archimedean absolute value, is (algebraically and topologically) isomorphic to either the real numbers or the complex numbers. This is sometimes also (confusingly) referred to as Ostrowski's theorem.

Pentagonal Number Theorem

In mathematics, the pentagonal number theorem, originally due to Euler, relates the product and series representations of the Euler function. It states that

$$\prod_{n=1}^{\infty}(1-x^n) = \sum_{k=-\infty}^{\infty}(-1)^k x^{k(3k-1)/2}.$$

In other words,

$$(1-x)(1-x^2)(1-x^3)\cdots = 1 - x - x^2 + x^5$$
$$+x^7 - x^{12} - x^{15} + x^{22} + x^{26} + \cdots.$$

A striking feature of this expansion is the amount of cancellation in the product. The indices 1, 2, 5, 7, 12, ... appearing on the right hand side are called pentagonal numbers (or more accurately, generalized pentagonal numbers).

If we treat this series as a power series, the radius of convergence is 1. One may also ignore issues of convergence, in which case the theorem holds as a statement about formal power series.

A Combinatorial Proof

The theorem can be given a combinatorial interpretation in terms of partitions. In particular, the left hand side is a generating function for the number of partitions of n into an even number of distinct parts minus the number of partitions of n into an odd number of distinct parts.

Each partition of n into an even number of distinct parts contributes +1 to the coefficient of x^n; each partition into an odd number of distinct parts contributes “1. (The article on unrestricted partition functions discusses this type of generating function.)

For example, the coefficient of x^5 is +1 because there are two ways to split 5 into an even number of distinct parts (4+1 and 3+2), but only one way to do so for an odd number of distinct parts (5 itself).

However, the coefficient of x^{12} is “1 because there are seven ways to partition 12 into an even number of distinct parts, but there are eight ways to partition 12 into an odd number of distinct parts. This interpretation leads to an elegant proof of the identity via involution.

Consider the Ferrers graph of any partition of n into distinct parts. For example, the diagram below shows $n = 20$ and the partition $20 = 7 + 6 + 4 + 3$.

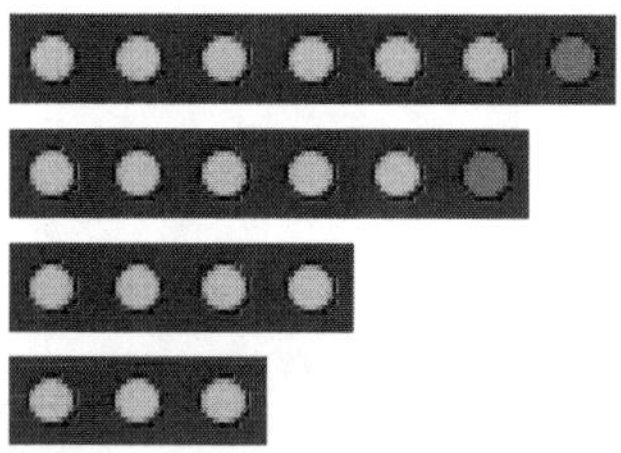

Let m be the number of elements in the smallest row of our graph. Let s be the number of elements in the rightmost 45 degree line of our graph (the dots which are red in the above diagram). In the diagram above, $m = 3$ and $s = 2$. If $m > s$, we take the rightmost 45 degree line and move it to form a new row, as in the graph below.

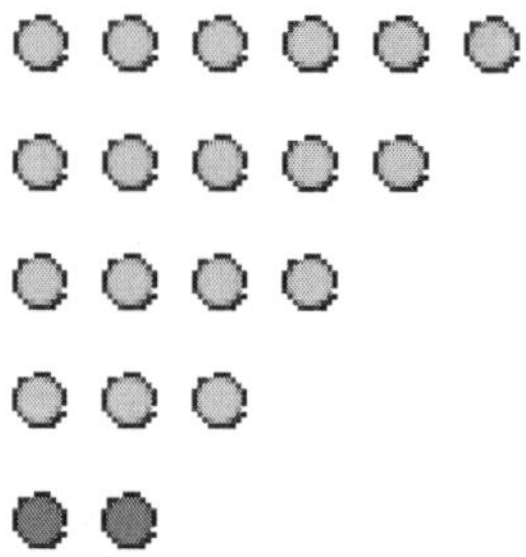

If this is not the case (as in our newly formed graph where $m = 2$, $s = 5$) we may reverse the process by moving the bottom row to form a new 45 degree line (in effect adding 1 element to each of the first m rows), taking us back into the first graph.

A bit of thought shows that applying the process twice will always bring us back to the original graph while also changing the parity of the number of rows.

This enables us to pair off the Ferrers' graphs of partitions contributing 1 and -1 to the original sum (resulting in no net coefficient value for a particular power of x) except in those cases where this operation cannot be performed. There are two such cases:

1) $m = s$ and the rightmost diagonal and bottom row meet. For example,

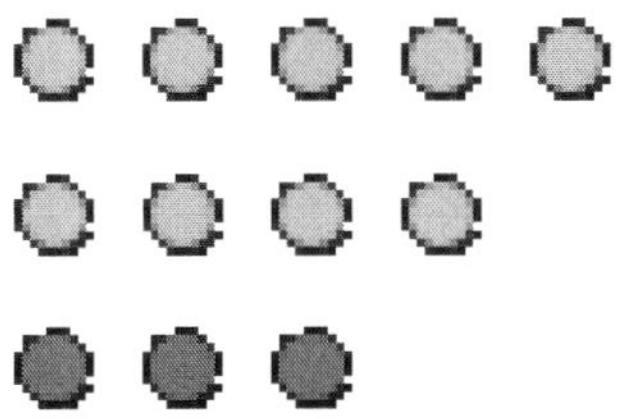

Attempting to perform the operation would lead us to:

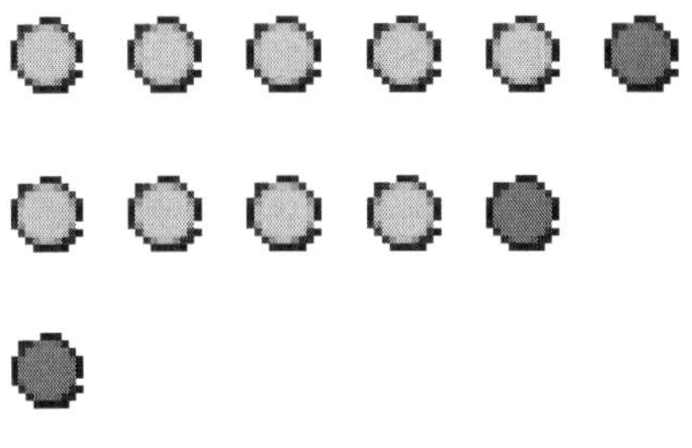

which fails to change the parity of the number of rows, and is not reversible in the sense that performing the operation again does *not* take us back to the original graph. If there are m elements in the last row of the original graph, then

$$n = m + (m+1) + (m+2) + \cdots + (2m-1) = \frac{m(3m-1)}{2} = \frac{k(3k-1)}{2}$$

where k is taken to equal m. Note that the factor associated with this partion is $(-1)^s$, which by construction equals $(-1)^m$ and $(-1)^k$.

2) $m = s+1$ and the rightmost diagonal and bottom row meet. For example,

Our operation requires us to move the right diagonal to the bottom row, but that would lead to two rows of 3 elements, forbidden since we're counting partitions into distinct parts. This is the previous case but with one fewer row, so

$$n = m + (m+1) + (m+2) + \cdots + (2m-2) = \frac{(m-1)(3m-2)}{2} = \frac{k(3k-1)}{2},$$

where m=1-k (which is negative). Here the associated factor is $(-1)^s$ with $s = m$-1 = -k, therefore the factor is also $(-1)^k$.

In summary, we have shown that partitions into an even number of distinct parts and an odd number of distinct parts exactly cancel each other out, except for pentagonal numbers, where there is exactly one case left over. But this is precisely what the right side of the identity says should happen, so we are finished.

This gives a beautiful recurrence for calculating , the number of partitions of n.

$$p_n = p_{n-1} + p_{n-2} - p_{n-5} - p_{n-7} + \cdots$$

or more formally,

$$p_n = \sum_i (-1)^{i-1} p_{n-q_i}$$

where the summation is over all (including negative) i and is the pentagonal number calculated as above.

Example Programme

Here is a simple Python programme which computes numbers of partitions using the Pentagonal number theorem.

```
pentagonal = lambda n : n*(3*n-1)/2

def generalised_pentagonal(n): # 0, 1, -1, 2, -2
    if n < 0: return 0
    if n%2 == 0: return pentagonal(n/2+1)
    else: return pentagonal(-(n/2+1))

pt = [1]
for n in range (1, 1000+1):
    r = 0
    f = -1
    i = 0
    while 1:
        k = generalised_pentagonal(i)
        if k > n:
            break
        if i%2==0: f = -f
        r += f*pt[n - k]
        i += 1
    pt.append(r)

print pt
```

Proizvolov's Identity

In mathematics, Proizvolov's identity is an identity concerning sums of differences of positive integers.

The identity was posed by Vyacheslav Proizvolov as a problem in the 1985 All-Union Soviet Student Olympiads (Savchev & Andreescu 2002, p. 66).

To state the identity, take the first $2N$ positive integers,

1, 2, 3, ..., $2N$ " 1, $2N$,

and partition them into two subsets of N numbers each. Arrange one subset in increasing order:

$$A_1 < A_2 < \cdots < A_N.$$

Arrange the other subset in decreasing order:

$$B_1 > B_2 > \cdots > B_N.$$

Then the sum

$$|A_1 - B_1| + |A_2 - B_2| + \cdots + |A_N - B_N|$$

is always equal to N^2.

Example

Take for example $N = 3$. The set of numbers is then {1, 2, 3, 4, 5, 6}. Select three numbers of this set, say 2, 3 and 5. Then the sequences A and B are:

$$A_1 = 2,\ A_2 = 3, \text{ and } A_3 = 5;$$
$$B_1 = 6,\ B_2 = 4, \text{ and } B_3 = 1.$$

The sum is

$$|A_1 - B_1| + |A_2 - B_2| + |A_3 - B_3| = |2-6| + |3-4| + |5-1| = 4+1+4 = 9,$$

which indeed equals 3^2.

Quadratic Reciprocity

In number theory, the law of quadratic reciprocity is a theorem about modular arithmetic which gives conditions for the solvability of quadratic equations modulo prime numbers.

There are a number of equivalent statements of the theorem, which consists of two "supplements" and the reciprocity law:

Let p, $q > 2$ be two distinct prime numbers. Then

(Supplement 1)

x^2 ≥ "1 (mod p) is solvable if and only if p a– 1 (mod 4).

(Supplement 2)

x^2 a– 2 (mod p) is solvable if and only if p a– ±1 (mod 8).

(Quadratic reciprocity)

Let $q^* = \pm q$ where the sign is plus if q a– 1 (mod 4) and minus if q a -1 (mod 4). (I.e. $|q^*| = q$ and q^* a– 1 (mod 4).) Then

x^2 a– p (mod q) is solvable if and only if x^2 a– q^* (mod p) is solvable.

Although the law can be used to tell whether any quadratic equation modulo a prime number has a solution, it does not provide any help at all for actually *finding* the solution. (The article on quadratic residues discusses algorithms for this.)

The theorem was conjectured by Euler and Legendre and first proven by Gauss. He refers to it as the "fundamental theorem" in the *Disquisitiones Arithmeticae* and his papers; privately he referred to it as the "golden theorem." He published six proofs, and two more were found in his posthumous papers. There are now over 200 published proofs.

The first section of this article does not use the Legendre symbol and gives the formulations of quadratic reciprocity found by Legendre and Gauss.

The Legendre-Jacobi symbol is introduced in the second section.

Terminology, Data, and Two Statements of the Theorem

A quadratic residue (mod n) is any number congruent to a square (mod n). A quadratic nonresidue (mod n) is any number which is not congruent to a square (mod n).

The adjective "quadratic" can be dropped if the context makes it clear that it is implied. When working modulo primes (as in this article), it is usual to treat zero as a special case. By doing so, the following statements become true:

Modulo a prime, there are an equal number of quadratic residues and nonresidues.

Modulo a prime, the product of two quadratic residues is a residue, the product of a residue and a nonresidue is a nonresidue, and the product of two nonresidues is a residue.

In this article, p and q always refer to distinct positive odd prime numbers.

–1 and the First Supplement

First of all, for which prime numbers is –1 a quadratic residue? Examining the table, we find –1 in rows 5, 13, 17, 29, 37, and 41 but not in rows 3, 7, 11, 19, 23, 31, 43 or 47.

(–1 a– 2 (mod 3), –1 a– 4 (mod 5), –1 a– 10 (mod 11), etc.)

The former primes are all a– 1 (mod 4), and the latter are all a– 3 (mod 4). This leads to

The first supplement to quadratic reciprocity:

The congruence $x^2 \equiv -1 \pmod{p}$

is solvable if and only if $p \equiv 1 \pmod{4}$.

±2 and the Second Supplement

For which prime numbers is 2 a quadratic residue? Examining the table, we find 2 in rows 7, 17, 23, 31, 41, and 47, but not in rows 3, 5, 11, 13, 19, 29, 37, or 43.

The former primes are all a– ±1 (mod 8), and the latter are all a– ±3 (mod 8). This leads to

The second supplement to quadratic reciprocity:

> The congruence $x^2 \equiv 2 \pmod{p}$
> is solvable if and only if $p \equiv \pm 1 \pmod{8}$.

–2 is in rows 3, 11, 17, 19, 41, 43, but not in rows 5, 7, 13, 23, 29, 31, 37, or 47. The former are a– 1 or a– 3 (mod 8), and the latter are a– 5 or a– 7 (mod 8).

±3

3 is in rows 11, 13, 23, 37, and 47, but not in rows 5, 7, 17, 19, 29, 31, 41, or 43.

The former are a– ±1 (mod 12) and the latter are all a– ±5 (mod 12).

–3 is in rows 7, 13, 19, 31, 37, and 43 but not in rows 5, 11, 17, 23, 29, 41, or 47. The former are a– 1 (mod 3) and the latter a– 2 (mod 3).

Since the only residue (mod 3) is 1, we see that –3 is a quadratic residue modulo every prime which is a residue (mod 3).

±5

5 is in rows 11, 19, 29, 31, and 41 but not in rows 3, 7, 13, 17, 23, 37, 43, or 47.

The former are ≡ ±1 (mod 5) and the latter are ≡ ±2 (mod 5).

Since the only residues (mod 5) are ±1, we see that 5 is a quadratic residue modulo every prime which is a residue (mod 5).

–5 is in rows 3, 7, 23, 29, 41, 43, and 47 but not in rows 11, 13, 17, 19, 31, or 37. The former are ≡ 1, 3, 7, 9 (mod 20) and the latter are ≡ 11, 13, 17, 19 (mod 20).

Gauss's Version

The observations aout –3 and +5 continue to hold: –7 is a residue (mod p) if and only if p is a residue (mod 7), –11 is a residue (mod p) if and only if p is a residue (mod 11), +13 is a residue (mod p) if and only if p is a residue (mod 13), ...

The more complicated-looking rules for the quadratic characters of +3 and –5, which depend upon congruences (mod 12) and (mod 20) respectively, are simply the ones for –3 and +5 working with the first supplement.

For example, for –5 to be a residue (mod p), either both 5 and –1 have to be residues (mod p) or they both have to be nonresidues: i.e., p has to be ≡ ±1 (mod 5) *and* a– 1 (mod 4), which is the same thing as p a– 1 or 9 (mod 20), or p has to be ≡ ±2 mod 5 *and* a– 3 mod 4, which is the same as p a– 3 or 7 (mod 20).

The generalization of the rules for –3 and +5 is Gauss's statement of quadratic reciprocity:

If $q \equiv 1 \pmod 4$ then

the congruence $x^2 \equiv p \pmod q$

is solvable if and only if $x^2 \equiv q \pmod p$ is, but

If $q \equiv 3 \pmod 4$ then

the congruence $x^2 \equiv p \pmod q$

is solvable if and only if $x^2 \equiv -q \pmod p$ is.

These statements may be combined:

Let $q^* = (\text{"}1)^{(q-1)/2}q$. Then the congruence x^2 a– p (mod q) is solvable if and only if x^2 a– q^* (mod p) is.

Legendre's Version

Another way to organize the data is to see which primes are residues mod which other primes, as illustrated in the above table. The entry in row p column q is R if q is a quadratic residue (mod p); if it is a nonresidue the entry is N.

If the row, or the column, or both, are a– 1 (mod 4) the entry is blue or green; if both row and column are a– 3 (mod 4), it is yellow or orange. The blue and green entries are symmetric around the diagonal: The entry for row p, column q is R (resp N) if and only if the entry at row q, column p, is R (resp N).

The yellow and orange ones, on the other hand, are antisymmetric: The entry for row p, column q is R (resp N) if and only if the entry at row q, column p, is N (resp R).

This observation is Legendre's statement of quadratic reciprocity:

If $p \equiv 1 \pmod 4$ or $q \equiv 1 \pmod 4$ (or both), then

$x^2 \equiv q \pmod p$ is solvable if and only if x^2

$\equiv p \pmod q$ is solvable.

If $p \equiv q \equiv 3 \pmod 4$, then

$x^2 \equiv q \pmod p$ is solvable if and only if x^2

$\equiv p \pmod q$ is not solvable.

It is a simple exercise to prove that Legendre's and Gauss's statements are equivalent - it requires no more than the first supplement and the facts about multiplying residues and nonresidues.

History and Alternative Statements

There are a number of ways to state the theorem. Keep in mind that Euler and Legendre did not have Gauss's congruence notation, nor did Gauss have the Legendre symbol.

In this article p and q always refer to distinct positive odd primes.

Fermat

Fermat proved (or claimed to have proved) a number of theorems about expressing a prime by a quadratic form:

$p = x^2 + y^2$ if and only if $p = 2$ or $p \equiv 1 \pmod 4$,

$p = x^2 + 2y^2$ if and only if $p = 2$ or $p \equiv 1, 3 \pmod 8$,

$p = x^2 + 3y^2$ if and only if $p = 3$ or $p \equiv 1 \pmod 3$.

He did not state the law of quadratic reciprocity, although the cases "1, ±2, and ±3 are easy deductions from these and other of his theorems.

He also claimed to have a proof that if the prime number p ends with 7, (in base 10) and the prime number q ends in 3, and p a– q a– 3 (mod 4), then

$$pq = x^2 + 5y^2.$$

Euler conjectured, and Lagrange proved, that

$$\text{If } \; p \equiv 1,9 \pmod{20} \text{ then } \; p = x^2 + 5y^2,$$

$$\text{if } p,q \equiv 3,7 \pmod{20} \text{ then } pq = x^2 + 5y^2.$$

Proving these and other statements of Fermat was one of the things that led mathematicians to the reciprocity theorem.

Euler

Translated into modern notation, Euler stated:

1) If q a– 1 (mod 4) then q is a quadratic residue (mod p) if and only if p a– r (mod q), where r is a quadratic residue of q.

2) If q a– 3 (mod 4) then q is a quadratic residue (mod p) if and only if p ≡ $\pm b^2$ (mod $4q$), where b is odd and not divisible by q.

This is equivalent to quadratic reciprocity.

He could not prove it, but he did prove the second supplement.

Legendre and His Symbol

Fermat proved that if p is a prime number and a is an integer,

$$a^p \equiv a \pmod{p}.$$

Thus, if p does not divide a,

$$a^{(p-1)/2} \equiv \pm 1 \pmod{p}.$$

Legendre lets a and A represent positive primes a– 1 (mod 4) and b and B positive primes a– 3 (mod 4), and sets out a table of eight theorems that together are equivalent to quadratic reciprocity:

Theorem	*When*	*it follows that*
I	$b^{(a-1)/2} \equiv +1 \pmod{a}$	$a^{(b-1)/2} \equiv +1 \pmod{b}$
II	$a^{(b-1)/2} \equiv -1 \pmod{b}$	$b^{(a-1)/2} \equiv -1 \pmod{a}$
III	$a^{(A-1)/2} \equiv +1 \pmod{A}$	$A^{(a-1)/2} \equiv +1 \pmod{a}$
IV	$a^{(A-1)/2} \equiv -1 \pmod{A}$	$A^{(a-1)/2} \equiv -1 \pmod{a}$
V	$a^{(b-1)/2} \equiv +1 \pmod{b}$	$b^{(a-1)/2} \equiv +1 \pmod{a}$
VI	$b^{(a-1)/2} \equiv -1 \pmod{a}$	$a^{(b-1)/2} \equiv -1 \pmod{b}$
VII	$b^{(B-1)/2} \equiv +1 \pmod{B}$	$B^{(b-1)/2} \equiv -1 \pmod{b}$
VIII	$b^{(B-1)/2} \equiv -1 \pmod{B}$	$B^{(b-1)/2} \equiv +1 \pmod{b}$

He says that since expressions of the form

$N^{(c-1)/2} \pmod{c}$ (where N and c are relatively prime) will come up so often he will abbreviate them as:

$$\left(\frac{N}{c}\right) = \pm 1 \equiv N^{(c-1)/2} \pmod{c}.$$

This is now known as the Legendre symbol, and an equivalent definition is used today: for all integers a and all odd primes p

$$\left(\frac{a}{p}\right) = \begin{cases} 0 \text{ if } a \equiv 0 \pmod{p} \\ +1 \text{ if } a \not\equiv 0 \pmod{p} \text{ and for some integer} \\ x,\ a \equiv x^2 \pmod{p} \\ -1 \text{ if there is no such } x. \end{cases}$$

Legendre's Version of Quadratic Reciprocity

$$\left(\frac{p}{q}\right) = \begin{cases} +\left(\frac{q}{p}\right) \text{ if } p \equiv 1 \pmod{4} \text{ or } q \equiv 1 \pmod{4} \\ -\left(\frac{q}{p}\right) \text{ if } p \equiv q \equiv 3 \pmod{4} \end{cases}$$

He notes that these can be combined:

$$\left(\frac{p}{q}\right)\left(\frac{q}{p}\right) = (-1)^{\frac{p-1}{2}\frac{q-1}{2}}.$$

A number of proofs, especially those based on Gauss's Lemma, explicitly calculate this formula.

Gauss

Gauss first proves the supplementary laws. He sets the basis for induction by proving the theorem for ±3 and ±5. Noting that it is easier to state for "3 and +5 than it is for +3 or –5, he states the general theorem in the form:

If *p* is a prime of the form 4*n* + 1 then *p*, but if *p* is of the form 4*n*+3 then –*p*, is a quadratic residue (resp. nonresidue) of every prime, which, with a positive sign, is a residue (resp. nonresidue) of *p*.

In the next sentence, he christens it the fundamental theorem (Gauss never used the word "reciprocity").

Introducing the notation *a* R *b* (resp. *a* N *b*) to mean *a* is a quadratic residue (resp. nonresidue) (mod *b*), and letting *a*, *a*2 , etc. represent positive primes a-1 (mod 4) and *b*, *b*2 , etc. positive primes a– 3 (mod 4), he breaks it out into the same 8 cases as Legendre:

Case	*If*	*Then*
1)	±*a* R *a*2	±*a*2 R *a*
2)	±*a* N *a*2	±*a*2 N *a*
3)	+*a* R *b*–*a* N *b*	±*b* R *a*
4)	+*a* N *b*–*a* R *b*	±*b* N *a*
5)	±*b* R *a*	+*a* R *b*–*a* N *b*
6)	±*b* N *a*	+*a* N *b*–*a* R *b*
7)	+*b* R *b*2 –*b* N *b*2	–*b*2 N *b*+*b*2 R *b*
8)	–*b* N *b*2 +*b* R *b*2	+*b*2 R *b*–*b*2 N *b*

In the next Article he generalizes this to what are basically the rules for the Jacobi symbol (below). Letting *A*, *A*2 , etc. represent any (prime or composite) positive numbers a– 1 (mod 4) and *B*, *B*2 , etc. positive numbers a– 3 (mod 4):

Case	*If*	*Then*
9)	±*a* R *A*	±*A* R *a*
10)	±*b* R *A*	+*A* R *b*–*A* N *b*
11)	+*a* R *B*	±*B* R *a*

12)	$-a$ R B	$\pm B$ N a
13)	$+b$ R B	$-B$ N b+N R b
14)	$-b$ R B	$+B$ R b–B N b

All of these cases take the form "if a prime is a residue (mod a composite), then the composite is a residue or nonresidue (mod the prime), depending on the congruences (mod 4)". He proves that these follow from cases 1) - 8).

Gauss needed, and was able to prove, a lemma similar to the one Legendre needed:

If $p \equiv 1 \pmod 8$ is prime, then there exists an odd prime $q < 2\sqrt{p} + 1$ such that $\left(\frac{p}{q}\right) = -1$.

The proof of quadratic reciprocity is by complete induction (i.e. assuming it is true for all numbers less than n allows the deduction it is true for n) for each of the cases 1) to 8).

Gauss's Version in Legendre Symbols

$$\left(\frac{p}{q}\right) = \begin{cases} \left(\frac{q}{p}\right) & \text{if } q \equiv 1 \pmod 4 \\ \left(\frac{-q}{p}\right) & \text{if } q \equiv 3 \pmod 4 \end{cases}$$

These can be combined:

Let $q^* = (-1)^{\frac{q-1}{2}} q$ (in other words $|q^*| = |q|$ and $q^* \equiv 1 \pmod 4$).

Then $\left(\frac{p}{q}\right) = \left(\frac{q^*}{p}\right)$.

A number of proofs of the theorem, especially those based on Gauss sums, or the splitting of primes in algebraic number fields, derive this formula.

Other Statements

Note that the statements in this section are equivalent to quadratic reciprocity: if, for example, Euler's version is assumed, the Legendre-Gauss version can be deduced from it, and vice-versa.

Euler

This form of quadratic reciprocity is derived from Euler's work:

$$\text{If } p \equiv \pm q \pmod{4a} \text{ then } \left(\frac{a}{p}\right) = \left(\frac{a}{q}\right).$$

Euler's statement can be proved by using Gauss's lemma.

Gauss

Gauss's fourth proof consists of proving this theorem (by comparing two formulas for the value of Gauss sums) and then restricting it to two primes:

Let *a*, *b*, *c*, ... be unequal positive odd primes, whose product is *n*, and let *m* be the number of them that are a= 3 (mod 4); check whether *n*/*a* is a residue of *a*, whether *n*/*b* is a residue of *b*, ... The number of nonresidues found will be even when *m* a– 0, 1 (mod 4), and it will be odd if *m* a – 2, 3 (mod 4).

He gives the example. Let $a = 3$, $b = 5$, $c = 7$, and $d = 11$. Three of these, 3, 7, and 11 ≡ 3 (mod 4), so $m \equiv 3 \pmod 4$.

5×7×11 R 3; 3×7×11 R 5; 3×5×11 R 7; and 3×5×7 N 11, so there are an odd number of nonresidues.

Eisenstein

Eisenstein formulates this:

$$\text{If } p \neq q, p' \neq q', p \equiv p' \pmod 4, \text{ and } q \equiv q' \pmod 4 \text{ then } \left(\frac{p}{q}\right)\left(\frac{q}{p}\right) = \left(\frac{p'}{q'}\right)\left(\frac{q'}{p'}\right).$$

Mordell

Mordell proved the following to be equivalent to quadratic reciprocity:

Let $a, b,$ and c be integers.

Then for every prime p that divides abc,

if $ax^2 + by^2 + cz^2 \equiv 0$
$\pmod{4abc/p}$ has a nontrivial solution

so does $ax^2 + by^2 + cz^2 \equiv 0 \pmod{4abc}$.

Jacobi Symbol

The Jacobi symbol is a generalization of the Legendre symbol; the main difference is that the bottom number has to be positive and odd, but does not have to be prime. If it is prime, the two symbols agree. It obeys the same rules of manipulation as the Legendre symbol. In particular

$$\left(\frac{-1}{n}\right) = (-1)^{(n-1)/2} = \begin{cases} 1 & \text{if } n \equiv 1 \pmod 4 \\ -1 & \text{if } n \equiv 3 \pmod 4 \end{cases}$$

$$\left(\frac{2}{n}\right) = (-1)^{(n^2-1)/8} = \begin{cases} 1 & \text{if } n \equiv 1 \text{ or } 7 \pmod 8 \\ -1 & \text{if } n \equiv 3 \text{ or } 5 \pmod 8 \end{cases}$$

and if both numbers are positive and odd (this is sometimes called "Jacobi's reciprocity law"):

$$\left(\frac{m}{n}\right)\left(\frac{n}{m}\right) = (-1)^{(m-1)(n-1)/4}.$$

However, if the Jacobi symbol is +1 and the bottom number is composite, it does not necessarily mean that the top number is a quadratic residue of the bottom one. Gauss's cases 9) - 14) above can be expressed in terms of Jacobi symbols:

$$\left(\frac{M}{p}\right) = (-1)^{(p-1)(M-1)/4}\left(\frac{p}{M}\right),$$

and since p is prime the left hand side is a Legendre symbol, and we know whether M is a residue (mod p) or not.

The formulas listed in the preceding section are true for Jacobi symbols as long as the symbols are defined. Euler's formula may be written

$$\left(\frac{a}{m}\right) = \left(\frac{a}{m \pm 4an}\right) \text{ where } n \text{ is an integer and } m \pm 4an > 0.$$

For example, $(\frac{2}{7})=(\frac{2}{15})=(\frac{2}{23})=(\frac{2}{31})\ldots=1$, and 2 is a residue mod the primes 7, 23 and 31: $3^2 \equiv 2 \pmod 7$, $5^2 \equiv 2 \pmod{23}$, and $8^2 \equiv 2 \pmod{31}$, but 2 is not a quadratic residue (mod 5), so it can't be one (mod 15). This is related to the problem Legendre had: if we know that $(\frac{a}{m})=-1$, we know that a is a nonresidue modulo every prime in the arithmetic series $m + 4a$, $m + 8a$, ..., if there *are* any primes in this series, but that wasn't proved until decades[25] after Legendre.

Eisenstein's formula requires relative primality conditions (which are true if the numbers are prime)

If a, b, a' and b' are positive and odd and

$\gcd(a,b)=\gcd(a',b')=1$, then

if $a \equiv a' \pmod 4$ and $b \equiv b'$

$\pmod 4$, $\left(\frac{a}{b}\right)\left(\frac{b}{a}\right)=\left(\frac{a'}{b'}\right)\left(\frac{b'}{a'}\right)$.

Other Rings

There are also quadratic reciprocity laws in rings other than the integers.

Higher Powers

The attempt to generalize quadratic reciprocity for powers higher than the second was one of the main goals that led 19th century mathematicians, including Carl Friedrich Gauss, J. P. G. Lejeune Dirichlet, Carl Gustav Jakob Jacobi, Gotthold Eisenstein, Richard Dedekind, Ernst Kummer, and David Hilbert to the study of general algebraic number fields and their rings of integers; specifically Kummer invented ideals in order to state and prove higher reciprocity laws.

The ninth in the list of 23 unsolved problems which David Hilbert proposed to the Congress of Mathematicians in 1900 asked for the "Proof of the most general reciprocity law [f]or an arbitrary number fiel≤. In 1923 Artin, building upon work by Furtwängler, Takagi, Hasse and others, discovered a general theorem for which all known reciprocity laws are special cases; he proved it in 1927. The links below provide more detailed discussions of these theorems.

Quartic Reciprocity

Quartic or biquadratic reciprocity is a collection of theorems in elementary and algebraic number theory that state conditions under which the congruence x^4 a- p (mod q) is solvable; the word "reciprocity" comes from the form of some of these theorems, in that they relate the solvability of the congruence x^4 a– p (mod q) to that of x^4 a– q (mod p).

History

Euler made the first conjectures about biquadratic reciprocity. Gauss published two monographs on biquadratic reciprocity. In the first one (1828) he proved Euler's conjecture about the biquadratic character of 2. In the second one (1832) he stated the biquadratic reciprocity law for the Gaussian integers and proved the supplementary formulas. He said that a third monograph would be forthcoming with the proof of the general theorem, but it never appeared. Jacobi presented proofs in his Königsberg lectures of 1836–37. The first published proofs were by Eisenstein.

Since then a number of other proofs of the classical (Gaussian) version have been found, as well as alternate statements. Lemmermeyer states that there has been an explosion of interest in the rational reciprocity laws since the 1970's.

Integers

A quartic or biquadratic residue (mod p) is any number congruent to the fourth power of an integer (mod p). If x^4 a– a (mod p) does not have an integer solution, a is a quartic or biquadratic nonresidue (mod p). As is often the case in number theory, it is easiest to work modulo prime numbers, so in this section all moduli p, q, etc., are assumed to positive, odd primes.

Gauss

The first thing to notice when working within the ring Z of integers is that if the prime number q is ≡ 3 (mod 4) then every quadratic residue (mod q) is also a biquadratic residue (mod q). The first supplement of quadratic reciprocity, states that "1 is a quadratic nonresidue (mod q), so that for any integer x, one of x and "x is a quadratic residue and the other one is a nonresidue.

Thus, if r a– a^2 (mod q) is a quadratic residue, then if a a– b^2 is a residue, r a– a^2 a– b^4 (mod q) is a biquadratic residue, and if a is a nonresidue, "a is a residue, "a a– b^2, and again, $r \equiv (-a)^2 \equiv b^4$ (mod q) is a biquadratic residue.

Therefore, the only interesting case is when the modulus p a– 1 (mod 4).

Gauss proved that if p a- 1 (mod 4) then the nonzero residue classes (mod p) can be divided into four sets, each containing (p–1)/4 numbers. Let e be a quadratic nonresidue. The first set is the quartic residues; the second one is e times the numbers in the first set, the third is e^2 times the numbers in the first set, and the fourth one is e^3 times the numbers in the first set. Another way to describe this division is to let g be a primitive root (mod p); then the first set is all the numbers whose indices with respect to this root are $\equiv 0$ (mod 4), the second set is all those whose indices are $\equiv 1$ (mod 4), etc. In the vocabulary of group theory, the first set is a subgroup of index 4 (of the multiplicative group $Z/pZ^{\times}$), and the other three are its cosets.

The first set is the biquadratic residues, the third set is the quadratic residues that are not quartic residues, and the second and fourth sets are the quadratic nonresidues. Gauss proved that "1 is a biquadratic residue if $p \equiv 1$ (mod 8) and a quadratic, but not biquadratic, residue, when $p \equiv 5$ (mod 8).

2 is a quadratic residue mod p if and only if $p \equiv \pm 1$ (mod 8). Since p is also $\equiv 1$ (mod 4), this means $p \equiv 1$ (mod 8). Every such prime is the sum of a square and twice a square.

Gauss proved

Let $q = a^2 + 2b^2 \equiv 1$ (mod 8) be a prime number. Then

2 is a biquadratic residue (mod q) if and only if $a \equiv \pm 1$ (mod 8), and

2 is a quadratic, but not a biquadratic, residue (mod q) if and only if $a \equiv \pm 3$ (mod 8).

Every prime $p \equiv 1$ (mod 4) is the sum of two squares. If $p = a^2 + b^2$ where a is odd and b is even, Gauss proved that

2 belongs to the first (respectively second, third, or fourth) class defined above if and only if $b \equiv 0$ (resp. 2, 4, or 6) (mod 8).

The first case of this is one of Euler's conjectures:

2 is a biquadratic residue of a prime $p \equiv 1 \pmod 4$ if and only if $p = a^2 + 64b^2$.

Gaussian Integers

Background

In his second monograph on biquadratic reciprocity Gauss displays some examples and makes conjectures that imply the theorems listed above for the biquadratic character of small primes. He makes some general remarks, and admits there is no obvious general rule at work. He goes on to say

The theorems on biquadratic residues gleam with the greatest simplilcity and genuine beauty only when the field of arithmetic is extended to imaginary numbers, so that without restriction, the numbers of the form $a + bi$ constitute the object of study ... we call such numbers integral complex numbers. [bold in the original]

These numbers are now called the ring of Gaussian integers, denoted by Z[i]. Note that i is a fourth root of 1.

In a footnote he adds

The theory of cubic residues must be based in a similar way on a consideration of numbers of the form $a + bh$ where h is an imaginary root of the equation $h^3 = 1$ ... and similarly the theory of residues of higher powers leads to the introduction of other imaginary quantities.

The numbers built up from a cube root of unity are now called the ring of Eisenstein integers.

The "other imaginary quantities" needed for the "theory of residues of higher powers" are the rings of integers of the cyclotomic number fields; the Gaussian and Eisenstein integers are the simplest examples of these.

Facts and Terminology

Gauss develops the arithmetic theory of the "integral complex numbers" and shows that it is quite similar to the arithmetic of ordinary integers. This is where the terms unit, associate, norm, and primary were introduced into mathematics.

The units are the numbers that divide 1. They are 1, *i*, “1, and “*i*. They are similar to 1 and “1 in the ordinary integers, in that they divide every number. The units are the powers of *i*.

Given a number λ = *a* + *bi*, its conjugate is *a* “ *bi* and its associates are the four numbers

λ = +*a* + *bi*

*i*λ = “*b* + *ai*

−λ = “*a* “ *bi*

“*i*λ = +*b* “ *ai*

The norm of λ = *a* + *bi* is the number Nλ = $a^2 + b^2$. If λ and μ are two Gaussian integers, Nλμ = Nλ Nμ; in other words, the norm is multiplicative. The norm of zero is zero, the norm of any other number is a positive integer. ε is a unit if and only if Nε = 1.

Gauss proves that Z[*i*] is a unique factorization domain and shows that the primes fall into three classes:

- 2 is a special case: 2 = $i^3 (1 + i)^2$. It is the only prime in Z divisible by the square of a prime in Z[*i*]. In algebraic number theory, 2 is said to ramify in Z[*i*].
- Positive primes in Z ≡ 3 (mod 4) are also primes in Z[*i*]. In algebraic number theory, these primes are said to remain inert in Z[*i*].
- Positive primes in Z ≡ 1 (mod 4) are the product of two conjugate primes in Z[*i*]. In algebraic number theory, these primes are said to split in Z[*i*].

Thus, inert primes are 3, 7, 11, 19, ... and a factorization of the split primes is

5 = (2 + *i*) × (2 “ *i*),

13 = (2 + 3*i*) × (2 “ 3*i*),

17 = (4 + *i*) × (4 “ *i*),

29 = (2 + 5*i*) × (2 “ 5*i*), ...

The associates and conjugate of a prime are also primes.

Note that the norm of an inert prime *q* is N*q* = q^2 ≡ 1 (mod 4); thus the norm of all primes other than 1 + *i* and its associates is ≡ 1 (mod 4).

Gauss calls a number in Z[i] odd if its norm is an odd integer. Thus all primes except 1 + i and its associates are odd.

The product of two odd numbers is odd and the conjugate and associates of an odd number are odd.

In order to state the unique factorization theorem, it is necessary to have a way of distinguishing one of the associates of a number. Gauss defines an odd number to be primary if it is ≡ 1 (mod $(1 + i)^3$).

It is straightforward to show that every odd number has exactly one primary associate. An odd number λ = $a + bi$ is primary if $a + b \equiv a$ “ $b \equiv 1$ (mod 4); i.e., $a \equiv 1$ and $b \equiv 0$, or $a \equiv 3$ and $b \equiv 2$ (mod 4). The product of two primary numbers is primary and the conjugate of a primary number is also primary.

The unique factorization theorem for Z[i] is: if λ ≠ 0, then

$$\lambda = \omega^{\mu}(1-\omega)^{\nu}\pi_1^{\alpha_1}\pi_2^{\alpha_2}\pi_3^{\alpha_3}\ldots$$

where 0 ≤ μ ≤ 3, ν e- 0, the π_is are primary primes and the α_is e- 1, and this representation is unique, up to the order of the factors.

The notions of congruence and greatest common divisor are defined the same way in Z[i] as they are for the ordinary integers Z. Because the units divide all numbers, a congruence (mod λ) is also true modulo any associate of λ, and any associate of a GCD is also a GCD.

Ribet's Theorem

In mathematics, Ribet's theorem (earlier called the epsilon conjecture or ε-conjecture) is a statement in number theory concerning properties of Galois representations associated with modular forms.

It was proposed by Jean-Pierre Serre and proved by Ken Ribet. The proof of epsilon conjecture was a significant step towards the proof of Fermat's Last Theorem.

As shown by Serre and Ribet, the Taniyama–Shimura conjecture (whose status was unresolved at the time) and the epsilon conjecture together imply that Fermat's Last Theorem is true.

Statement

Let E be an elliptic curve with integer coefficients in a Neron minimal model. Suppose that the discriminant Δ of E is written as the product $\Pi p^{\delta}{}_{p}$ of prime powers $p^{\delta}{}_{p}$ and similarly the conductor N of E is the product $\Pi p^{n}{}_{p}$ of prime powers. Suppose that E is a modular elliptic curve, then we can perform a level descent modulo primes –! dividing one of the exponents δ_p of a prime dividing the discriminant.

If $p^{\delta}{}_{p}$ is an odd prime power factor of Δ and if p divides N only once (i.e. n_p=1), then there exists another elliptic curve E', with conductor $N' = N/p$, such that the coefficients of the L-series of E are congruent modulo –! to the coefficients of the L-series of E'

.

The epsilon conjecture is a *relative* statement: assuming that a given elliptic curve E over Q is modular, it predicts the precise level of E.

History

In his thesis, Yves Hellegouarch came up with the idea of associating solutions (a,b,c) of Fermat's equation with a completely different mathematical object: an elliptic curve.[citation needed]. If –! is an odd prime and a, b, and c are positive integers such that

$$a^{\ell} + b^{\ell} = c^{\ell},$$

then a corresponding Frey curve is an algebraic curve given by the equation

$$y^2 = x(x - a^{\ell})(x + b^{\ell}).$$

This is a nonsingular algebraic curve of genus one defined over Q, and its projective completion is an elliptic curve over Q. In 1982 Gerhard Frey called attention to the unusual properties of the same curve as Hellegouarch, now called a Frey curve.

This provided a bridge between Fermat and Taniyama by showing that a counterexample to Fermat's Last Theorem would create such a curve that would not be modular. The conjecture attracted considerable interest when Frey (1986) suggested that the Taniyama–Shimura–Weil conjecture implies Fermat's Last Theorem. However, his argument was not complete. In 1985, Jean-

Pierre Serre proposed that a Frey curve could not be modular and provided a partial proof of this. This showed that a proof of the semistable case of the Taniyama-Shimura conjecture would imply Fermat's Last Theorem. Serre did not provide a complete proof and what was missing became known as the epsilon conjecture or ε-conjecture. In the summer of 1986, Ribet (1990) proved the epsilon conjecture, thereby proving that the Taniyama–Shimura–Weil conjecture implied Fermat's Last Theorem.

Implication of Fermat's Last Theorem

Suppose that the Fermat equation with exponent –! e– 3 had a solution in non-zero integers *a, b, c*. Let us form the corresponding Frey curve *E*. It is an elliptic curve and one can show that its discriminant Δ is equal to 16 $(abc)^{2-!}$ and its conductor *N* is the radical of *abc,* i.e. the product of all distinct primes dividing *abc*. By the Taniyama–Shimura conjecture, *E* is a modular elliptic curve. Since *N* is square-free, by the epsilon conjecture one can perform level descent modulo –!. Repeating this procedure, we will eliminate all odd primes from the conductor and reach the modular curve $X_0(2)$ of level 2. However, this curve is not an elliptic curve since it has genus zero, resulting in a contradiction.

Riemann–Siegel Formula

In mathematics, the Riemann–Siegel formula is an asymptotic formula for the error of the approximate functional equation of the Riemann zeta function, an approximation of the zeta function by a sum of two finite Dirichlet series.

It was found by Siegel (1932) in unpublished manuscripts of Bernhard Riemann dating from the 1850s. Siegel derived it from the Riemann–Siegel integral formula, an expression for the zeta function involving contour integrals.

It is often used to compute values of the Riemann–Siegel formula, sometimes in combination with the Odlyzko–Schönhage algorithm which speeds it up considerably. When used along the critical line, it is often useful to use it in a form where it becomes a formula for the Z function.

If *M* and *N* are non-negative integers, then the zeta function is equal to

$$\zeta(s)=\sum_{n=1}^{N}\frac{1}{n^{s}}+\gamma(1-s)\sum_{n=1}^{M}\frac{1}{n^{1-s}}+R(s)$$

where

$$\gamma(s)=\pi^{1/2-s}\Gamma(s/2)/\Gamma((1-s)/2)$$

is the factor appearing in the functional equation $\zeta(s) = \gamma(s)\ \zeta(1-s)$, and

$$R(s)=\frac{-\Gamma(1-s)}{2\pi i}\int\frac{(-x)^{s-1}e^{-Nx}dx}{e^{x}-1}$$

is a contour integral whose contour starts and ends at +– and circles the singularities of absolute value at most $2\pi M$. The approximate functional equation gives an estimate for the size of the error term. Siegel (1932) and Edwards (1974) derive the Riemann–Siegel formula from this by applying the method of steepest descent to this integral to give an asymptotic expansion for the error term $R(s)$ as a series of negative powers of Im(s). In applications s is usually on the critical line, and the positive integers M and N are chosen to be about $(2\pi\ \mathrm{Im}(s))^{1/2}$. Gabcke (1979) found good bounds for the error of the Riemann–Siegel formula.

Riemann's Integral Formula

Riemann showed that

$$\int_{0\searrow 1}\frac{e^{-i\pi u^{2}+2\pi ipu}du}{e^{\pi iu}-e^{-\pi iu}}=\frac{e^{i\pi p^{2}}-e^{i\pi p}}{e^{i\pi p}-e^{-i\pi p}}$$

where the contour of integration is a line of slope "1 passing between 0 and 1 (Edwards 1974, 7.9).

He used this to give the following integral formula for the zeta function:

$$\pi^{-s/2}\Gamma(s/2)\zeta(s)=$$

$$\pi^{-s/2}\Gamma(s/2)\int_{0\swarrow 1}\frac{x^{-s}e^{\pi ix^{2}}}{e^{\pi ix}-e^{-\pi ix}}$$

$$dx+\pi^{-(1-s)/2}\Gamma((1-s)/2)\int_{0\searrow 1}\frac{x^{s-1}e^{-\pi ix^{2}}}{e^{\pi ix}-e^{-\pi ix}}dx$$

Riemann–von Mangoldt Formula

In mathematics, the Riemann–von Mangoldt formula, named for Bernhard Riemann and Hans Carl Friedrich von Mangoldt, states that the number $N(T)$ of zeros of the Riemann zeta function with imaginary part greater than 0 and less than or equal to T satisfies

$$N(T) = \frac{T}{2\pi}\log\frac{T}{2\pi} - \frac{T}{2\pi} + O(\log T).$$

The formula was stated by Riemann in his famous paper *On the Number of Primes Less Than a Given Magnitude* (1859) and proved by von Mangoldt in 1905.

Serre's Modularity Conjecture

In mathematics, Serre's modularity conjecture, introduced by Serre (1975, 1987) based on some 1973–1974 correspondence with John Tate, states that an odd irreducible two-dimensional Galois representation over a finite field arises from a modular form, and a stronger version of his conjecture specifies the weight and level of the modular form.

Formulation

The conjecture concerns the absolute Galois group $G_{\mathbb{Q}}$ of the rational number field $\mathbb{Q}$.

Let ρ be an absolutely irreducible, continuous, two-dimensional representation of $G_{\mathbb{Q}}$ over a finite field that is odd (meaning that complex conjugation has determinant 1)

$$F = \mathbb{F}_{\ell^r}$$

of characteristic ℓ,

$$\rho : G_{\mathbb{Q}} \to \mathrm{GL}_2(F).$$

To any normalized modular eigenform

$$f = q + a_2q^2 + a_3q^3 + \cdots$$

of level $N = N(\rho)$, weight $k = k(\rho)$, and some Nebentype character

$$\chi : \mathbb{Z}/N\mathbb{Z} \rightarrow F^{*} \ ,$$

a theorem due to Shimura, Deligne, and Serre-Deligne attaches to a representation

$$\rho_f : G_{\mathbb{Q}} \rightarrow \mathrm{GL}_2(\mathcal{O}),$$

where $\mathcal{O}$ is the ring of integers in a finite extension of $\mathbb{Q}_\ell$. This representation is characterized by the condition that for all prime numbers p, coprime to we have $N\ell$

$$\mathrm{Trace}(\rho_f(\mathrm{Frob}_p)) = a_p$$

and

$$\det(\rho_f(\mathrm{Frob}_p)) = p^{k-1}\chi(p).$$

Reducing this representation modulo the maximal ideal of gives a mod representation of .

Serre's conjecture asserts that for any as above, there is a modular eigenform such that

$$\overline{\rho_f} \cong \rho \, .$$

The level and weight of the conjectural form are explicitly calculated in Serre's article. In addition, he derives a number of results from this conjecture, among them Fermat's Last Theorem and the now-proven Taniyama-Weil (or Taniyama-Shimura) conjecture, now known as the Modularity Theorem (although this implies Fermat's Last Theorem, Serre proves it directly from his conjecture).

Optimal Level and Weight

The strong form of Serre's conjecture describes the level and weight of the modular form.

The optimal level is the Artin conductor of the representation, with the power of l removed.

Proof

A proof of the level 1 and small weight cases of the conjecture was obtained during 2004 by Chandrashekhar Khare and Jean-Pierre Wintenberger,[1] and by Luis Dieulefait, independently. In 2005, Chandrashekhar Khare obtained a proof of the level 1 case of Serre conjecture, and in 2008 a proof of the full conjecture in collaboration with Jean-Pierre Wintenberger.

Siegel's Theorem on Integral Points

In mathematics, Siegel's theorem on integral points is the 1929 result of Carl Ludwig Siegel, that for a smooth algebraic curve C of genus g defined over a number field K, presented in affine space in a given coordinate system, there are only finitely many points on C with coordinates in the ring of integers O of K, provided $g > 0$. This result covers the Mordell curve, for example.

This was proved by combining a version of the Thue–Siegel–Roth theorem, from diophantine approximation, with the Mordell–Weil theorem from diophantine geometry (required in Weil's version, to apply to the Jacobian variety of C). It was the first major result on diophantine equations that depended only on the genus, not any special algebraic form of the equations. For $g > 1$ it was in the end superseded by Faltings' theorem.

Siegel's result was ineffective, since Thue's method in diophantine approximation also is ineffective in describing possible very good rational approximations to algebraic numbers. Effective results in some cases derive from Baker's method.

Six Exponentials Theorem

In mathematics, specifically transcendental number theory, the six exponentials theorem is a result that, given the right conditions on the exponents, guarantees the transcendence of at least one of a set of exponentials.

Statement

If $x_1, x_2, \ldots, x_d$ are d complex numbers that are linearly independent over the rational numbers, and $y_1, y_2, \ldots, y_l$ are l complex numbers that are also linearly independent over the rational numbers, and if $dl > d + l$, then at least one of the following dl numbers is transcendental:

$$\exp(x_i y_j), \quad (1 \le i \le d, 1 \le j \le l).$$

The most interesting case is when $d = 3$ and $l = 2$, in which case there are six exponentials, hence the name of the result. The theorem is weaker than the related but thus far unproved four exponentials conjecture, whereby the strict inequality $dl > d + l$ is replaced with dl e– $d + l$, thus allowing $d = l = 2$.

The theorem can be stated in terms of logarithms by introducing the set L of logarithms of algebraic numbers:

$$\mathcal{L} = \{\lambda \in \mathbb{C}: e^{\lambda} \in \overline{\mathbb{Q}}\}.$$

The theorem then says that if λ_{ij} are elements of L for $i = 1, 2$ and $j = 1, 2, 3$, such that λ_{11}, λ_{12}, and λ_{13} are linearly independent over the rational numbers, and λ_{11} and λ_{21} are also linearly independent over the rational numbers, then the matrix

$$M = \begin{pmatrix} \lambda_{11} & \lambda_{12} & \lambda_{13} \\ \lambda_{21} & \lambda_{22} & \lambda_{23} \end{pmatrix}$$

has rank 2.

History

A special case of the result where x_1, x_2, and x_3 are logarithms of positive integers, $y_1 = 1$, and y_2 is real, was first mentioned in a paper by Leonidas Alaoglu and Paul Erdõs from 1944 in which they try to prove that the ratio of consecutive colossally abundant numbers is always prime. They claimed that Carl Ludwig Siegel knew of a proof of this special case, but it is not recorded. Using the special case they manage to prove that the ratio of consecutive colossally abundant numbers is always either a prime or a semiprime.

The theorem was first explicitly stated and proved in its complete form independently by Serge Lang and Kanakanahalli Ramachandra in the 1960s.

Five Exponentials Theorem

A stronger, related result is the five exponentials theorem, which is as follows. Let x_1, x_2 and y_1, y_2 be two pairs of complex numbers, with each pair being linearly independent over the rational numbers, and let γ be a non-zero algebraic number. Then at least one of the following five numbers is transcendental:

$$e^{x_1 y_1}, e^{x_1 y_2}, e^{x_2 y_1}, e^{x_2 y_2}, e^{\gamma x_2 / x_1}.$$

This theorem implies the six exponentials theorem and in turn is implied by the as yet unproven four exponentials conjecture, which says that in fact one of the first four numbers on this list must be transcendental.

Sharp six exponentials theorem

Another related result that implies both the six exponentials theorem and the five exponentials theorem is the sharp six exponentials theorem.

This theorem is as follows. Let x_1, x_2, and x_3 be complex numbers that are linearly independent over the rational numbers, and let y_1 and y_2 be a pair of complex numbers that are linearly independent over the rational numbers, and suppose that β_{ij} are six algebraic numbers for $1 \le i \le 3$ and $1 \le j \le 2$ such that the following six numbers are algebraic:

$$e^{x_1 y_1 - \beta_{11}}, e^{x_1 y_2 - \beta_{12}}, e^{x_2 y_1 - \beta_{21}}, e^{x_2 y_2 - \beta_{22}}, e^{x_3 y_1 - \beta_{31}}, e^{x_3 y_2 - \beta_{32}}.$$

Then $x_i y_j = \beta_{ij}$ for $1 \le i \le 3$ and $1 \le j \le 2$. The six exponentials theorem then follows by setting $\beta_{ij} = 0$ for every i and j, while the five exponentials theorem follows by setting $x_3 = \gamma/x_1$ and using Baker's theorem to ensure that the x_i are linearly independent.

There is a sharp version of the five exponentials theorem as well, although it as yet unproven so is known as the sharp five exponentials conjecture.

This conjecture implies both the sharp six exponentials theorem and the five exponentials theorem, and is stated as follows.

Let x_1, x_2 and y_1, y_2 be two pairs of complex numbers, with each pair being linearly independent over the rational numbers, and let α, β_{11}, β_{12}, β_{21}, β_{22}, and γ be six algebraic numbers with $\gamma \ne 0$ such that the following five numbers are algebraic:

$$e^{x_1 y_1 - \beta_{11}}, e^{x_1 y_2 - \beta_{12}}, e^{x_2 y_1 - \beta_{21}}, e^{x_2 y_2 - \beta_{22}}, e^{(\gamma x_2 / x_1) - \alpha}.$$

Then $x_i y_j = \beta_{ij}$ for $1 \le i, j \le 2$ and $\gamma x_2 = \alpha x_1$.

A consequence of this conjecture that isn't currently known would be the transcendence of e^{π^2}, by setting $x_1 = y_1 = \beta_{11} = 1$, $x_2 = y_2 = i\pi$, and all the other values in the statement to be zero.

Strong Six Exponentials Theorem

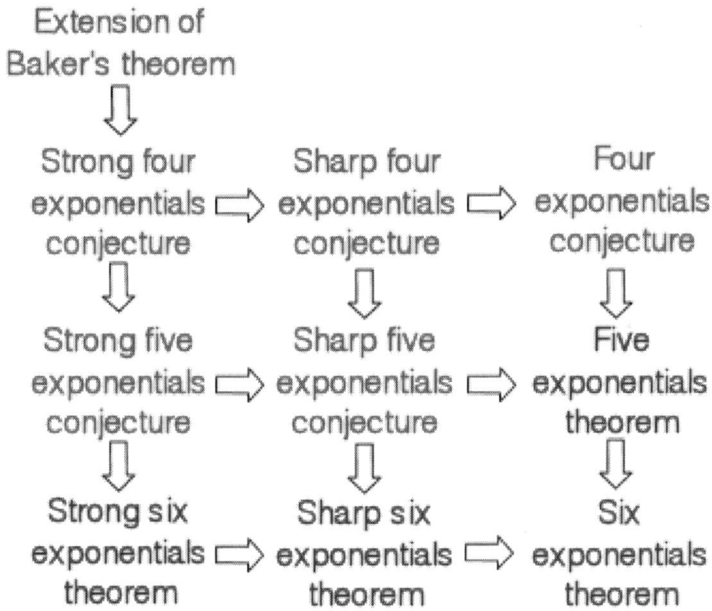

The logical implications between the various problems in this circle. Those in red are as yet unproven while those in blue are known results.

The top most result refers to that discussed at Baker's theorem, while the four exponentials conjectures are detailed at the four exponentials conjecture article.

A further strengthening of the theorems and conjectures in this area are the strong versions. The strong six exponentials theorem is a result proved by Damien Roy that implies the sharp six exponentials theorem. This result concerns the vector space over the algebraic numbers generated by 1 and all logarithms of algebraic numbers, denoted here as L''. So L'' is the set of all complex numbers of the form

$$\beta_0 + \sum_{i=1}^{n} \beta_i \log \alpha_i,$$

for some n e– 0, where all the β_i and α_i are algebraic and every branch of the logarithm is considered. The strong six exponentials theorem then says that if x_1, x_2, and x_3 are complex numbers that are linearly independent over the algebraic numbers, and if y_1 and y_2 are a pair of complex numbers that are also linearly independent over the algebraic numbers then at least one of the six numbers $x_i y_j$ for $1 \leq i \leq 3$ and $1 \leq j \leq 2$ is not in L''. This is stronger than the standard six exponentials theorem which says that one of these six numbers is not simply the logarithm of an algebraic number.

There is also a strong five exponentials conjecture that would imply both the strong six exponentials theorem and the sharp five exponentials conjecture. This conjecture claims that if x_1, x_2 and y_1, y_2 are two pairs of complex numbers, with each pair being linearly independent over the algebraic numbers, then at least one of the following five numbers is not in L'':

$$x_1y_1, x_1y_2, x_2y_1, x_2y_2, x_1 / x_2.$$

All the above conjectures and theorems are consequences of the unproven extension of Baker's theorem, that logarithms of algebraic numbers that are linearly independent over the rational numbers are automatically algebraically independent too. The diagram on the right shows the logical implications between all these results.

Skolem–Mahler–Lech Theorem

In additive number theory, the Skolem–Mahler–Lech theorem, named after Thoralf Skolem, Kurt Mahler, and Christer Lech, states that the indices of the null elements of a linear recurrence sequence are the union of a finite set and finitely many arithmetic progressions. The proofs of this result use p-adic analysis.

Sophie Germain's Theorem

In number theory, Sophie Germain's theorem is a statement about the divisibility of solutions to the equation $x^p + y^p = z^p$ of Fermat's Last Theorem. Specifically, Sophie Germain proved that the product xyz must be divisible by p^2 if an auxiliary prime θ can be found such that two conditions are satisfied:

1. No two p^{th} powers differ by one modulo θ; and
2. p is itself not a pth power modulo θ.

Conversely, the first case of Fermat's Last Theorem (the case in which p does not divide xyz) must hold for every prime p for which even one auxiliary prime can be found. Germain identified such an auxiliary prime θ for every prime less than 100. The theorem and its application to primes p less than 100 were attributed to Germain by Adrien-Marie Legendre in 1823.

Stickelberger's Theorem

In mathematics, Stickelberger's theorem is a result of algebraic number theory, which gives some information about the Galois module structure of class groups of cyclotomic fields. A special case was first proven by Ernst Kummer (1847) while the general result is due to Ludwig Stickelberger (1890).

The Stickelberger Element and the Stickelberger Ideal

Let K_m denote the mth cyclotomic field, i.e. the extension of the rational numbers obtained by adjoining the mth roots of unity to Q (where m e– 2 is an integer). It is a Galois extension of Q with Galois group G_m isomorphic to the multiplicative group of integers modulo m $(Z/mZ)^\times$. The Stickelberger element (of level m or of K_m) is an element in the group ring $Q[G_m]$ and the Stickelberger ideal (of level m or of K_m) is an ideal in the group ring $Z[G_m]$. They are defined as follows. Let ζ_m denote a primitive mth root of unity. The isomorphism from $(Z/mZ)^\times$ to G_m is given by sending a to σ_a defined by the relation

$$\sigma_a(\zeta) = \zeta\, a$$

$$m\,.$$

The Stickelberger element of level m is defined as

$$\theta(K_m) = \frac{1}{m} \sum_{\substack{a=1 \\ (a,m)=1}}^{m} a\cdot\sigma_a^{-1} \in \mathbf{Q}[G_m].$$

The Stickelberger ideal of level m, denoted $I(K_m)$, is the set of integral multiples of $\theta(K_m)$ which have integral coefficients, i.e.

$$I(K_m) = \theta(K_m)\mathbf{Z}[G_m] \cap \mathbf{Z}[G_m].$$

More generally, if F be any abelian number field whose Galois group over Q is denoted G_F, then the Stickelberger element of F

and the Stickelberger ideal of F can be defined. By the Kronecker–Weber theorem there is an integer m such that F is contained in K_m. Fix the least such m (this is the (finite part of the) conductor of F over Q).

There is a natural group homomorphism $G_m \to G_F$ given by restriction, i.e. if σ " G_m, its image in G_F is its restriction to F denoted $\text{res}_m\sigma$.

The Stickelberger element of F is then defined as

$$\theta(F) = \frac{1}{m} \sum_{\substack{a=1 \\ (a,m)=1}}^{m} a \cdot \text{res}_m \sigma_a^{-1} \in \mathbf{Q}[G_F].$$

The Stickelberger ideal of F, denoted $I(F)$, is defined as in the case of K_m, i.e.

$$I(F) = \theta(F)\mathbf{Z}[G_F] \cap \mathbf{Z}[G_F].$$

In the special case where $F = K_m$, the Stickelberger ideal $I(K_m)$ is generated by (a " σ_a)θ(K_m) as a varies over Z/mZ. This not true for general F.

Examples

- If F is a totally real field of conductor m, then

$$\theta(F) = \frac{\phi(m)}{2[F:\mathbf{Q}]} \sum_{\sigma \in G_F} \sigma,$$

where φ is the Euler totient function and $[F : Q]$ is the degree of F over Q.

Statement of the Theorem

Stickelberger's Theorem

Let F be an abelian number field. Then, the Stickelberger ideal of F annihilates the class group of F.

Note that θ(F) itself need not be an annihilator, but any multiple of it in $Z[G_F]$ is.

Explicitly, the theorem is saying that if α " $Z[G_F]$ is such that and if J is any fractional ideal of F, then is a principal ideal.

Størmer's Theorem

In number theory, Størmer's theorem, named after Carl Størmer, gives a finite bound on the number of consecutive pairs of smooth numbers that exist, for a given degree of smoothness, and provides a method for finding all such pairs using Pell equations. It follows from the Thue–Siegel–Roth theorem that there are only a finite number of pairs of this type, but Størmer gave a procedure for finding them all. Louis Mordell wrote about this result, saying that it "is very pretty, and there are many applications of it."

Application in Music Theory

In the theory of musical tuning, musical tones can be described as integer multiples of a fundamental frequency, and the multiples generated by products of small prime numbers are of particular importance: in Pythagorean tuning, only tones corresponding to integer multiples of the form $2^i \times 3^j$ are allowed, while in just tuning, only the tones corresponding to numbers of the form $2^i \times 3^j \times 5^k$ are allowed, where i, j, and k may range over any non-negative integer value. The difference between one tone and another forms a musical interval that can be measured by the ratio between the two corresponding integers, and in music the superparticular ratios between consecutive integers are of particular importance.

Størmer's theorem implies that, for Pythagorean tuning, the only possible superparticular ratios are 2/1 (the octave), 3/2 (the perfect fifth), 4/3 (the perfect fourth), and 9/8 (the whole step). That is, the only pairs of consecutive integers that have only powers of two and three in their prime factorizations are (1,2), (2,3), (3,4), and (8,9). For just tuning, six additional superparticular ratios are available: 5/4, 6/5, 10/9, 16/15, 25/24, and 81/80; all are musically meaningful.

Some modern musical theorists have developed p-limit musical tuning systems for primes p larger than 5; Størmer's theorem applies as well in these cases, and describes how to calculate the set of possible superparticular ratios for these systems.

Formal Statement of the Theorem

Formally, the theorem states that, if one chooses a finite set $P = \{p_1, \ldots p_k\}$ of prime numbers and considers the set of integers

$$S = \left\{ p_1^{e_1} p_2^{e_2} \cdots p_k^{e_k} \mid e_i \in \{0,1,2,\ldots\} \right\}$$

that can be generated by products of numbers in P, then there are only finitely many pairs of consecutive numbers in S. Further, it gives a method of finding them all using Pell equations.

The Procedure

Størmer's original procedure involves solving a set of roughly 3^k Pell equations, in each one finding only the smallest solution. A simplified version of the procedure, due to D. H. Lehmer,[4] is described below; it solves fewer equations but finds more solutions in each equation.

Let P be the given set of primes, and define a number to be P-smooth if all its prime factors belong to P. Assume $p_1 = 2$; otherwise there can be no consecutive P-smooth numbers. Lehmer's method involves solving the Pell equation

$$x^2 - 2qy^2 = 1$$

for each P-smooth square-free number q other than 2. Each such number q is generated as a product of a subset of P, so there are 2^k-1 Pell equations to solve. For each such equation, let x_i, y_i be the generated solutions, for i in the range [1,max(3,(p_k+1)/2)], where p_k is the largest of the primes in P.

Then, as Lehmer shows, all consecutive pairs of P-smooth numbers are of the form (x_i - 1)/2, (x_i + 1)/2. Thus one can find all such pairs by testing the numbers of this form for P-smoothness.

Example

To find the ten consecutive pairs of {2,3,5}-smooth numbers giving the superparticular ratios for just tuning, let P = {2,3,5}. There are seven P-smooth squarefree numbers q (omitting the eighth P-smooth squarefree number, 2): 1, 3, 5, 6, 10, 15, and 30, each of which leads to a Pell equation. The number of solutions per Pell equation required by Lehmer's method is max(3,(5+1)/2) = 3, so this method generates three solutions to each Pell equation, as follows.

- For q = 1, the first three solutions to the Pell equation $x^2 - 2y^2 = 1$ are (3,2), (17,12), and (99,70). Thus, for each of the three values x_i = 3, 17, and 99, Lehmer's method tests the pair $(x_i-1)/2,(x_i+1)/2$ for smoothness; the three pairs to be tested are (1,2), (8,9), and (49,50). Both (1,2) and (8,9) are pairs of consecutive *P*-smooth numbers, but (49,50) is not, as 49 has 7 as a prime factor.

- For q = 3, the first three solutions to the Pell equation $x^2 - 6y^2 = 1$ are (5,2), (49,20), and (485,198). From the three values x_i = 5, 49, and 485 Lehmer's method forms the three candidate pairs of consecutive numbers $(x_i-1)/2,(x_i+1)/2$: (3,2), (25,24), and (243,242). Of these, (3,2) and (25,24) are pairs of consecutive *P*-smooth numbers but (243,242) is not.
- For q = 5, the first three solutions to the Pell equation x^2 " $10y^2 = 1$ are (19,6), (721,228), and (27379,8658). The Pell solution (19,6) leads to the pair of consecutive *P*-smooth numbers (9,10); the other two solutions to the Pell equation do not lead to *P*-smooth pairs.
- For q = 6, the first three solutions to the Pell equation x^2 " $12y^2 = 1$ are (7,2), (97,28), and (1351,390). The Pell solution (7,2) leads to the pair of consecutive *P*-smooth numbers (3,4).
- For q = 10, the first three solutions to the Pell equation x^2 " $20y^2 = 1$ are (9,2), (161,36), and (2889,646). The Pell solution (9,2) leads to the pair of consecutive *P*-smooth numbers (4,5) and the Pell solution (161,36) leads to the pair of consecutive *P*-smooth numbers (80,81).
- For q = 15, the first three solutions to the Pell equation x^2 " $30y^2 = 1$ are (11,2), (241,44), and (5291,966). The Pell solution (11,2) leads to the pair of consecutive *P*-smooth numbers (5,6).
- For q = 30, the first three solutions to the Pell equation x^2 " $60y^2 = 1$ are (31,4), (1921,248), and (119071,15372). The Pell solution (31,4) leads to the pair of consecutive *P*-smooth numbers (15,16).

Counting solutions

Størmer's original result can be used to show that the number of consecutive pairs of integers that are smooth with respect to a set of k primes is at most $3^k - 2^k$. Lehmer's result produces a tighter bound for sets of small primes: $(2^k - 1) \times \max(3,(p_k+1)/2)$.

The number of consecutive pairs of integers that are smooth with respect to the first k primes are

1, 4, 10, 23, 40, 68, 108, 167, 241, 345, ... (sequence A002071 in OEIS).

The largest integer from all these pairs, for each k, is

2, 9, 81, 4375, 9801, 123201, 336141, 11859211, ... (sequence A117581 in OEIS).

OEIS also lists the number of pairs of this type where the larger of the two integers in the pair is square (sequence A117582 in OEIS) or triangular (sequence A117583 in OEIS), as both types of pair arise frequently.

Generalizations and Applications

Chein (1976) used Størmer's method to prove Catalan's conjecture on the nonexistence of consecutive perfect powers (other than 8,9) in the case where one of the two powers is a square.

Mabkhout (1993) proved that every number $x^4 + 1$, for $x > 3$, has a prime factor greater than or equal to 137. Størmer's theorem is an important part of his proof, in which he reduces the problem to the solution of 128 Pell equations.

Several authors have extended Størmer's work by providing methods for listing the solutions to more general diophantine equations, or by providing more general divisibility criteria for the solutions to Pell equations.

Subspace Theorem

In mathematics, the subspace theorem is a result obtained by Wolfgang M. Schmidt (1972). It states that if $L_1,\ldots,L_n$ are linearly independent linear forms in n variables with algebraic coefficients and if $\varepsilon>0$ is any given real number, then the non-zero integer points x with

$$|L_1(x)\cdots L_n(x)|<|x|^{-\epsilon}$$

lie in a finite number of proper subspaces of Q^n. A quantitative form of the theorem, in which the number of subspaces containing all solutions, was also obtained by Schmidt, and the theorem was generalised by Schlickewei (1977) to allow more general absolute values on number fields.

The theorem may be used to obtain results on Diophantine equations such as Siegel's theorem on integral points and solution of the S-unit equation.

A Corollary on Diophantine Approximation

The following corollary to the subspace theorem is often itself referred to as the *subspace theorem*. If $a_1,...,a_n$ are algebraic such that $1,a_1,...,a_n$ are linearly independent over Q and ε>0 is any given real number, then there are only finitely many rational n-tuples $(x_1/y,...,x_n/y)$ with

$$| a_i - x_i / y |< y^{-(1+1/n+\epsilon)}, \quad i = 1,\dots,n.$$

The specialization $n = 1$ gives the Thue–Siegel–Roth theorem. One may also note that the exponent $1+1/n+\varepsilon$ is best possible by Dirichlet's theorem on diophantine approximation.

Szemerédi's Theorem

In number theory, Szemerédi's theorem is a result that was formerly the Erdõs–Turán conjecture (not to be confused with Erdõs–Turán conjecture on additive bases). In 1936 Erdõs and Turán conjectured for every value d called density $0 < d < 1$ and every integer k there is a number $N(d, k)$ such that every subset A of $\{1, ..., N\}$ of cardinality dN contains a length-k arithmetic progression, provided $N > N(d, k)$.

This generalizes the statement of van der Waerden's theorem.

The cases $k = 1$ and $k = 2$ are trivial. The case $k = 3$ was established in 1953 by Klaus Roth by an adaptation of the Hardy–Littlewood circle method. The case $k = 4$ was established in 1969 by Endre Szemerédi by a combinatorial method. Using an approach similar to the one he used for the case $k = 3$, Roth gave a second proof for this in 1972.

Finally, the case of general k was settled in 1975, also by Szemerédi, by an ingenious and complicated extension of the previous combinatorial argument ("a masterpiece of combinatorial reasoning", R. L. Graham). Several further proofs are now known, the most important amongst them being those by Hillel Fürstenberg in 1977, using ergodic theory, and by Timothy Gowers in 2001, using both Fourier analysis and combinatorics.

Let k be a positive integer and let $0 < \delta \leq 1/2$. A finitary version of the theorem states that there exists a positive integer

$$N = N(k, \delta)$$

such that every subset of {1, 2, ..., N} of size at least δN contains an arithmetic progression of length k. The best known bounds for $N(k, \delta)$ are

$$C^{\log(1/\delta)^{k-1}} \leq N(k,\delta) \leq 2^{2^{\delta^{-2^{2^{k+9}}}}}$$

with $C > 1$. The lower bound is due to Behrend (for $k = 3$) and Rankin, and the upper bound is due to Gowers. When $k = 3$ better upper bounds are known; Bourgain has proved that

$$N(3,\delta) \leq C^{\delta^{-2}\log(1/\delta)},$$

In 2010, Sanders has proved that a subset of {1, 2, ..., N} with no arithmetic subsequence of length three is of size .

Thue Equation

In mathematics, a Thue equation is a Diophantine equation of the form

$f(x,y) = r,$

where f is an irreducible bivariate form of degree at least 3 over the rational numbers, and r is a nonzero rational number. It is named after Axel Thue who in 1909 proved a theorem, now called Thue's theorem, that Thue equation has finitely many solutions in integers x and y. The Thue equation is soluble effectively: there is an explicit bound on the solutions x, y of the form where constants C_1 and C_2 depend only on the form f. A stronger result holds, that if K is the field generated by the roots of f then the equation has only finitely many solutions with x and y integers of $K$and again these may be effectively determined.

Solving Thue Equations

Solving a Thue equation can be described as an algorithm ready for implementation in software. In particular, it is implemented in the following computer algebra systems:

- in PARI/GP as functions *thueinit()* and *thue()*.
- in Magma computer algebra system as functions *ThueObject()* and *ThueSolve()*.

Tijdeman's Theorem

In number theory, Tijdeman's theorem states that there are at most a finite number of consecutive powers.

Stated another way, the set of solutions in integers x, y, n, m of the exponential diophantine equation

$$y^m = x^n + 1,$$

for exponents n and m greater than one, is finite.

The theorem was proven by Dutch number theorist Robert Tijdeman in 1976, and provided a strong impetus towards the eventual proof of Catalan's conjecture by Preda Mihãilescu.

Mihãilescu's theorem states that there is only one member to the set of consecutive power pairs, namely 9=8+1.

That the powers are consecutive is essential to Tijdeman's proof; if we replace a difference of one by any other difference k and ask for the number of solutions of

$$y^m = x^n + k$$

with n and m greater than one we have an unsolved problem. It is conjectured that this set also will be finite; its finiteness would follow, for instance, from the abc conjecture.

Tunnell's theorem

In number theory, Tunnell's theorem gives a partial resolution to the congruent number problem, and under the Birch and Swinnerton-Dyer conjecture, a full resolution.

The congruent number problem asks which rational numbers can be the area of a right triangle with all three sides rational. Tunnell's theorem relates this to the number of integral solutions to a few fairly simple Diophantine equations.

The theorem is named for Jerrold B. Tunnell, a number theorist at Rutgers University, who proved it in 1983.

both in , then there exist x and y in Z/pZ such that $x^3 = \alpha$ and $y^3 = \alpha + 1$; consequently, $x^3 + 1 = y^3$. Therefore is the number of solutions to $x^3 + 1 = y^3$ in the finite field Z/pZ. The other coefficients have similar interpretations. Gauss's determination of the coefficients of the products of the periods therefore counts the number of points on these elliptic curves, and as a byproduct he proves the analog of the Riemann hypothesis.

The Weil conjectures in the special case of algebraic curves were conjectured by Artin (1924). The case of curves over finite fields was proved by Weil, finishing the project started by Hasse's theorem on elliptic curves over finite fields. Their interest was obvious enough from within number theory: they implied upper bounds for exponential sums, a basic concern in analytic number theory.(Moreno 2001)

What was really eye-catching, from the point of view of other mathematical areas, was the proposed connection with algebraic topology. Given that finite fields are *discrete* in nature, and topology speaks only about the *continuous*, the detailed formulation of Weil (based on working out some examples) was striking and novel. It suggested that geometry over finite fields should fit into well-known patterns relating to Betti numbers, the Lefschetz fixed-point theorem and so on.

The analogy with topology suggested that a new homological theory be set up applying within algebraic geometry. This took two decades (it was a central aim of the work and school of Alexander Grothendieck) building up on initial suggestions from Serre. The rationality part of the conjectures was proved first by Bernard Dwork (1960), using p-adic methods. Grothendieck (1965) and his collaborators established the rationality conjecture, the functional equation and the link to Betti numbers by using the properties of étale cohomology, a new cohomology theory developed by Grothendieck and Artin for attacking the Weil conjectures, as outlined in Grothendieck (1960). Of the four conjectures the analogue of the Riemann hypothesis was the hardest to prove. Motivated by the proof of Serre (1960) of an analogue of the Weil conjectures for Kähler manifolds, Grothendieck envisioned a proof based on his standard conjectures on algebraic cycles (Kleiman 1968).

However, the standard conjectures remain open (except for the hard Lefschetz theorem, which was proved by Deligne by extending his work on the Weil conjectures), and the analogue of the Riemann hypothesis was proved by Deligne (1974), using the étale cohomology theory but circumventing the use of standard conjectures by an ingenious argument.

Deligne (1980) found and proved a generalization of the Weil conjectures, bounding the weights of the pushforward of a sheaf.

Examples

The projective line:

The simplest example (other than a point) is to take X to be the projective line. The number of points of X over a field with q^m elements is just $N_m = q^m + 1$ (where the "+ 1– comes from the "point at infinity"). The zeta function is just

$1/(1 - q^{-s})(1 - q^{1-s})$.

It is easy to check all parts of the Weil conjectures directly. For example, the corresponding complex variety is the Riemann sphere and its initial Betti numbers are 1, 0, 1.

Projective Space

It is not much harder to do n dimensional projective space. The number of points of X over a field with q^m elements is just $N_m = 1 + q^m + q^{2m} + ... + q^{nm}$. The zeta function is just

$1/(1 - q^{-s})(1 - q^{1-s})(1 - q^{2-s}) ... (1 - q^{n-s})$.

It is again easy to check all parts of the Weil conjectures directly. (Complex projective space gives the relevant Betti numbers, which nearly determine the answer.)

The number of points on the projective line and projective space are so easy talculate because they can be written as disjoint unions of a finite number of copies of affine spaces. It is also easy to prove the Weil conjectures for other spaces, such as Grassmannians and flag varieties, which have the same "paving" property.

Elliptic Curves

These give the first non-trivial cases of the Weil conjectures (proved by Hasse). If E is an elliptic curve over a finite field with projective line P^1, with a finite number of singular fibres with very mild (quadratic) singularities. The theory of monodromy of Lefschetz pencils, introduced for complex varieties (and ordinary cohomology) by Lefschetz (1924), and extended by Grothendieck (1972) and Deligne & Katz (1973) to l-adic cohomology, relates the cohomology of V to that of its fibres. The relation depends on the space E_x of vanishing cycles, the subspace of the cohomology $H^{\leq 1}(V_x)$ of a non-singular fibre V_x, spanned by classes that vanish on singular fibres.

- The Leray spectral sequence relates the middle cohomology group of V to the cohomology of the fibre and base. The hard part to deal with is more or less a group $H^1(P^1, j_*E)$ = $H1$

c(U,E), where U is the points the projective line with non-singular fibres, and j is the inclusion of U into the projective line, and E is the sheaf with fibres the spaces E_x of vanishing cycles.

Deligne's Second Proof

Deligne (1980) found and proved a generalization of the Weil conjectures, bounding the weights of the pushforward of a sheaf. In practice it is this generalization rather than the original Weil conjectures that is mostly used in applications, such as the hard Lefschetz theorem. Much of the second proof is a rearrangement of the ideas of his first proof. The main extra idea needed is an argument closely related to the theorem of Hadamard and de la Vallée Poussin, used by Deligne to show that various L-series do not have zeros with real part 1.

A constructible sheaf on a variety over a finite field is called pure of weight β if for all points x the eigenvalues of the Frobenius at x all have absolute value $N(x)^{\beta/2}$, and is called mixed of weight $\leq\beta$ if it can be written as repeated extensions by pure sheaves with weights $\leq\beta$.

Deligne's theorem states that if f is a morphism of schemes of finite type over a finite field, then $R^if_!$ takes mixed sheaves of weight $\leq\beta$ to mixed sheaves of weight $\leq\beta$+i. The original Weil conjectures follow by taking f to be a morphism from a smooth projective variety to a point and considering the constant sheaf Q_l on the variety. This gives an upper bound on the absolute values of the eigenvalues of Frobenius, and Poincaré duality then shows that this is also a lower bound.

In general $R^i f_!$ does not take pure sheaves to pure sheaves. However it does when a suitable form of Poincaré duality holds, for example if *f* is smooth and proper, or if one works with perverse sheaves rather than sheaves as in Beilinson, Bernstein & Deligne (1982).

Inspired by the work of Witten (1982) on Morse theory, Laumon (1987) found another proof, using Deligne's *l*-adic Fourier transform, which allowed him to simplify Deligne's proof by avoiding the use of the method of Hadamard and de la Vallée Poussin. His proof generalizes the classical calculation of the absolute value of Gauss sums using the fact that the norm of a Fourier transform has a simple relation to the norm of the original function.

Kiehl & Weissauer (2001) used Laumon's proof as the basis for their exposition of Deligne's theorem. Katz (2001) gave a further simplification of Laumon's proof, using monodromy in the spirit of Deligne's first proof. Kedlaya (2006) gave another proof using the Fourier transform, replacing etale cohomology with rigid cohomology.

Applications

- Deligne (1980) was able to prove the hard Lefschetz theorems (part of Grothendieck's standard conjectures) using his second proof of the Weil conjectures.
- Deligne (1971) had previously shown that the Ramanujan-Petersson conjecture follows from the Weil conjectures.
- Deligne used the Weil conjectures to prove estimates for exponential sums.

Zeckendorf's Theorem

The first 160 integers (on the X-axis) broken down into Zeckendorf representation. Each rectangle's colour corresponds with a Fibonacci number and its height corresponds with the number's value.

Zsigmondy's Theorem

In number theory, Zsigmondy's theorem, named after Karl Zsigmondy, states that if $a > b > 0$ are coprime integers, then for any natural number $n > 1$ there is a prime number p (called a *primitive prime divisor*) that divides $a^n - b^n$ and does not divide $a^k - b^k$ for any positive integer $k < n$, with the following exceptions:

- $a = 2$, $b = 1$, and $n = 6$; or
- $a + b$ is a power of two, and $n = 2$.

This generalized Bang's theorem which states that if $n>1$ and n is not equal to 6, then 2^n-1 has a prime divisor not dividing any 2^k-1 with $k<n$.

Similarly $a^n + b^n$, has at least one primitive prime divisor with the exception $2^3 + 1^3 = 9$

Zsigmondy's theorem is often useful, especially in group theory, where it is used to prove that various groups have distinct orders except when they are known to be the same

History

The mathematical theorem was discovered by Zsigmondy working in Vienna from 1894 until 1925.

Bibliography

Apostol, Tom M.: *An Introduction to the Theory of Numbers,* Springer, New York, 2010.

Clark, Walter Eugene: *The Âryabhamîya of Âryabhama: An ancient Indian work on mathematics and astronomy.* University of Chicago Press, 2000.

Davenport, Harold; Montgomery, Hugh L.: *Multiplicative number theory,* Springer, New York, 2000.

Edwards, Harold M.: *Fermat's Last Theorem: a genetic introduction to algebraic number theory,* Springer, New York, 2000.

Friberg, Jöran: *Historia Mathematica,* Cambridge Univ. Press, Cambridge, 2011.

Gauss, Carl Friedrich; Waterhouse, William C.: *Disquisitiones Arithmeticae.* Springer, New York, 2006.

Goldfeld, Dorian M.: *Elementary proof of the prime number theorem,* Cambridge Univ. Press, Cambridge, 2003.

Granville, Andrew: *Analytic number theory,* Princeton University Press, USA, 2008.

Guthrie, Kenneth Sylvan: *The Pythagorean Sourcebook and Library,* Phanes Press, Michigan, 2007.

Hardy, Godfrey Harold; Wright, E. M.: *An introduction to the theory of numbers* , Oxford University Press, Oxford, 2008.

Heath, Thomas L.: *A History of Greek Mathematics,* Oxford Clarendon Press, Oxford, 2001.

Huffman, Carl A.: *Stanford Encyclopaedia of Philosophy,* Platonist Press, New Jersey, 2011.

Iwaniec, Henryk; Kowalski, Emmanuel: *Analytic number theory,* American Mathematical Society Colloquium Publications, America, 2004.

Lam, Lay Yong; Ang, Tian Se: *Fleeting Footsteps: Tracing the conception of arithmetic and algebra in ancient China,* World Scientific, Singapore, 2004.

Mahoney, M. S.: *The mathematical career of Pierre de Fermat,* Princeton University Press, Singapore, 1994.

Neugebauer, Otto E.: *The exact sciences in antiquity,* Dover Publications, New York, 2009.

Plofker, Kim: *Mathematics in India,* Princeton University Press, Singapore, 2008.

Porphyry; Guthrie, K. S.: *Life of Pythagoras,* Platonist Press, New Jersey, 1999.

Sachau, Eduard: *Alberuni's India: An account of the religion,* Kegan, Paul, Trench, Trübner & Co.., London, 2008.

Smith, D. E.: *History of Mathematics,* Dover Publications, New York, 2008.

van der Waerden, Bartel L.; Dresden, Arnold: *Science Awakening,* Oxford University Press, New York, 2011.

Varadarajan, V. S.: *Euler through time: a new look at old themes,* Dover Publications, New York, 2006.

von Fritz, Kurt : *Classics in the History of Greek Mathematics,* Springer, New York, 2004.

Index

❑❑❑